How Funding Instability Affects Army Programs

T0306331

David Kassing, R. William Thomas,
Frank Camm, Carolyn Wong

Prepared for the United States Army
Approved for public release; distribution unlimited

RAND ARROYO CENTER

The research described in this report was sponsored by the United States Army under Contract No. DASW01-01-C-0003.

Library of Congress Cataloging-in-Publication Data

How funding instability affects Army programs / David Kassing ... [et al.].
p. cm.
Includes bibliographical references.
ISBN 978-0-8330-4177-7 (pbk. : alk. paper)
1. United States. Army—Appropriations and expenditures. 2. United States. Air Force—Appropriations and expenditures. 3. Program budgeting—United States. I. Kassing, David.

UA25.5.H78 2007
355.6'2260973—dc22

2007019723

Published 2007 by the RAND Corporation
1776 Main Street, P.O. Box 2138, Santa Monica, CA 90407-2138
1200 South Hayes Street, Arlington, VA 22202-5050
4570 Fifth Avenue, Suite 600, Pittsburgh, PA 15213-2665
RAND URL: http://www.rand.org/
To order RAND documents or to obtain additional information, contact
Distribution Services: Telephone: (310) 451-7002;
Fax: (310) 451-6915; Email: order@rand.org

Preface

In 2003, the Assistant Secretary of the Army for Acquisition, Logistics, and Technology (ASA[ALT]) and the Deputy Chief of Staff, G-8, U.S. Army, sponsored a project that asked how changes in funding for Army weapon system programs affect the management success of those programs. This monograph is the product of that project. The analysis described addresses Army project funding instability and its relationship to acquisition program performance; it also provides insights based on case studies of three Army programs that experienced high funding instability.

This monograph should interest senior managers and analysts responsible for weapon system programs both in the Army and in the Department of Defense more broadly, as well as the defense financial management community and decisionmakers responsible for designing and overseeing policy relevant to these programs.

This project was conducted in RAND Arroyo Center's Strategy, Doctrine, and Resources Program. RAND Arroyo Center, part of the RAND Corporation, is a federally funded research and development center sponsored by the United States Army.

The Project Unique Identification Code (PUIC) for the project that produced this document is DAPRRX009.

Contents

Figures

Tables

Summary

At any given time, the U.S. Army is supporting many large and small weapon and equipment programs at every stage of the acquisition cycle, from concept development to disposition. And each year, senior-level decisionmakers make changes, both large and small, to the funding allocations for these individual programs for a host of reasons, such as shifts in priorities and requirements, the emergence of new opportunities, engineering-design modifications, contractor-performance and technical problems, and overall budget reductions. After the fact, the decisions to make these funding changes may turn out to have been mistakes; but they nevertheless reflect the best judgment of Army and Office of the Secretary of Defense (OSD) leaders given the information available to them at the time. However, regardless of the root causes or reasons for the changes, the result is the same: funding instability that programs must absorb.

Some Army officials are concerned that funding changes in Army weapon programs may occur without adequate attention being paid to how they will affect the management of those programs. Professional acquisition officials understand that funding instability affects program management in general, but those who are responsible for initiating the funding changes may not have this understanding. Moreover, circumstances in individual programs differ so much that even knowledgeable senior officials may not fully appreciate how funding changes are likely to affect the management of a particular program. As a result, funding decisions made during program reviews may result in unintended effects on a program's performance goals, cost, or schedule.

This project sought empirical information from the experience of recent Army weapon system programs to clarify the effects of changes in their funding. Three different approaches were used to shed light on the research issues. One was an exploratory quantitative analysis to define and measure funding instability and determine whether it was associated with symptoms of program management problems. Eighteen major Army programs were used for this analysis. The second approach employed three case studies of the activities of individual programs to determine whether funding instability occurred and, if so, how it occurred and what its adverse effects were. The third approach was an analysis of evidence on funding instability in Army and Air Force programs since 2000.

Evidence from Exploratory Quantitative Analyses

Cost growth and schedule slippage have been persistent problems in Department of Defense (DoD) and Army programs. The literature on acquisition management shows that external direction leads to slippage in acquisition program schedules and that technical complexity is a major factor in program cost growth. Moreover, program stability tends to limit cost growth. Conversely, instability in programs creates these two adverse effects through changes in quantities and in the productivity of existing plants and equipment, as well as through subtle changes in management and subcontractor activities. These effects differ among programs, and the literature detects no strong patterns across all the services and DoD agencies.

An earlier comparison of adverse effects in Army and other service acquisition programs generally confirmed the lack of strong distinctions. Development cost growth in all services increases with time, but the average cost growth is quite similar in the Army's and other services' major programs. The general pattern of procurement cost growth is also similar across the services, although it is somewhat higher for the Army. Schedule slippage is large in both the Army and the other services. In all these comparisons, however, the variations within all services dominate the differences in service averages.

Our analysis of data from Army acquisition programs (see Table S.1 for a list of the 18 programs selected) provided new measures of funding instability. These measures compare the absolute value of differences between planned and actual funding (expressed in constant dollars) for a five-year period with the planned funding level. In effect, they summarize the differences between the actual funding profiles and those estimated at Milestone B (i.e., the point at which the decision to begin system development is made). The measure is the ratio of the absolute value of changes summed over five years to the sum of the planned funding for the same five years. Thus, funding instability is relative to the funding planned at Milestone B. Since the relevant milestone decisions occurred in different years, the funding instability measures reflect data from different periods. In short, the measure of funding instability is the difference between planned and actual funding, and the higher this measure, or "score," the greater the instability. Thus, the funding instability scores are based on planning estimates made relatively early in the program, a time of substantial uncertainties about technologies, contractors, and costs. Our use of the new measures revealed wide variation among Army programs in both development and procurement funding instability.

In general, funding instability is higher for procurement than for development—more than twice as large on average. Furthermore, changes between planned and actual funding are the norm when year-by-year comparisons are made. It is important, however, to keep in mind that funding instability can be either a cause or an effect of program problems.

Our analysis also estimated adverse outcomes, such as cost growth and schedule changes, for the same 18 Army programs. Although these measures of adverse outcomes also showed wide variations among the 18 programs, a statistical analysis found only one association between funding instability and adverse outcomes: Procurement funding instability is correlated with schedule slippage. Such an association may or may not indicate a cause-and-effect relationship.

In sum, cost growth and schedule slippage in Army programs are generally comparable to those experienced in other services, and funding instability in Army programs is not strongly associated with

Table S.1
Funding Instability Scores for 18 Selected Army Programs

System	Type	Development Funding Instability Score	Procurement Funding Instability Score
Family of Medium Tactical Vehicles (FMTV)	Vehicle	27.9	52.5
Bradley Fighting Vehicle System Upgrade	Vehicle	36.7	32.2
Black Hawk Utility Helicopter (UH-60A/L)	Helicopter	2.3	32.0
Longbow Apache Airframe (AFM)	Helicopter	3.5	24.3
Chinook Improved Cargo Helicopter (CH-47F)	Helicopter	58.6	31.1
Javelin	Missile	86.2	75.4
Guided Multiple Launch Rocket System (GMLRS)	Missile	39.0	68.6
Longbow Hellfire	Missile	22.1	41.3
Brilliant Anti-Armor Technology Preplanned Product Improvement (BAT P3I)	Munition	44.1	100.0
Sense and Destroy Armor Submunition (SADARM 155)	Munition	nil	88.4
Advanced Threat Infrared Countermeasure/Common Missile Warning System (ATIRCM/CMWS)	Electronic	19.6	95.3
Forward Area Air Defense Command, Control, and Intelligence (FAAD C2I)	Electronic	44.7	99.9
Longbow Apache Fire Control Radar (FCR)	Electronic	2.0	18.2
Joint Surveillance Target Attack Radar System Ground Station Module (JSTARS GSM)	Electronic	32.2	33.3
Secure Mobile Anti-Jam Reliable Tactical Terminal (SMART-T)	Electronic	20.2	66.7
Joint Surveillance Target Attack Radar System Common Ground Station (JSTARS CGS)	Electronic	18.6	92.7
AN/TYQ-45 Maneuver Control System (MCS)	Electronic	6.5	62.4
Combat Service Support Control Systems (CSSCS)	Electronic	10.7	76.8
Average:		26.4	60.6

SOURCE: Calculated from *Selected Acquisition Report* data.

these two adverse outcomes. The only way to determine whether there is a connection is to conduct a more detailed analysis. The three case studies we conducted, which are summarized next, are a step in this direction.

Evidence from Case Studies

Three case studies are not sufficient for drawing strong policy conclusions. Nonetheless, the cases studied suggest that the major sources of funding instability originated outside the Army, in events such as the Cold War's end and the Global War on Terrorism. The Army's establishment of ambitious program goals also contributed to high levels of funding instability. But funding changes made during top-level internal Army reviews did not appear to create significant difficulties for Army program managers in our three case studies.

The Javelin program's experiences show these effects. The program approved for development of the Javelin missile system in 1989 was recognized as ambitious at the time. Technical problems followed, and the development schedule had to be extended, resulting in what was high development funding instability by our measure. In addition, before the Javelin could move into production, the Cold War ended, Army forces were cut, and the Javelin procurement objectives were cut nearly in half. These "fact of life" changes led to high procurement funding instability. Their effects included development cost growth, an extended development schedule, and substantial increases in procurement unit acquisition costs.

A similar mix of internal and external sources of funding instability was uncovered in our case study of the FAAD C2I program, which was approved for development in 1986. The complexity and ambitiousness of the original program goals led to problems that were accommodated by several program restructurings. For example, delays in the deliveries of government furnished equipment led to slips in related activities. A solicitation for an ambitious subsystem failed to attract a single qualified bidder. And FAAD C2I requirements for identification friend or foe capabilities were shifted to the Air Force. These and other events created substantial development funding instability. They also led to essentially all of the planned procurement slipping out of the initial five-year period. The end of the Cold War and the lessons drawn from Operation Desert Storm led to additional restructuring, as well as to reductions in procurement reflecting reductions in the Army's force structure. These changes caused a five-year slip in the program's completion of system development and demonstration. However, force structure reductions ended up countering unit cost increases enough to make the FAAD C2I program's cost growth less than the average cost growth among the 18 programs included in the quantitative analysis.

The evidence from the CH-47F case study reveals that the program's funding instability was the result of recommendations made by the program manager and the initial effects of the Global War on Terrorism. Faced with the contractor's substantially increased procurement unit costs, the CH-47F program manager recommended a one-year slip in the start of low-rate production and an associated one-year extension of the development effort. Army leadership approved this recommendation. Then, as a result of operations in Afghanistan, a decision was reached in 2003 to reorder the planned production, moving acquisition of the special-operations version of the CH-47 from later in the production run to its start. The extension of the development effort caused the schedule to slip by eight months but had only a small effect on development costs. The reordering of the production program, however, led to a substantial slip in the schedule for achieving the "first unit equipped" with CH-47Fs.

None of the problems that were revealed in the case studies as significant for Army program managers originated in across-the-board or targeted funding cuts initiated by Army leadership. This is not to say that funding instability did not create problems for program managers. Contractual, program, budget, and public information activities all required adaptations done in close coordination. But the root causes of funding instability were for the most part grounded in significant events beyond the Army leaders' control.

Evidence from More-Recent Experience

Since the concerns of Army leaders about the effects of funding changes on program management are more likely to have arisen from more-recent program experiences, we performed additional analyses of funding instability from 2000 through 2004.

The research and development (R&D) funding instability experienced by Army programs during FY 2000 through FY 2004 was considerably greater than that experienced by Army programs in the 1980s and 1990s. This higher degree of instability may well be the result of Army leaders' efforts to obtain the funds needed to create new transformational programs by modifying or canceling ongoing programs. Army programs also experienced greater R&D funding instability than did Air Force programs in this period.

This was not the case for procurement funding instability during the period. In this case, the Army and Air Force programs experienced a roughly similar degree of instability. And the procurement instability for the Army programs in 2000 through 2004 was considerably less than it had been for Army programs in the 1980s and 1990s. One possible explanation is that the set of programs that had passed Milestone C (i.e., the start of procurement) and entered the production phase by 2000 were in consonance with Army leaders' intermediate goals of maintaining and modernizing the force while pursuing transformation with newer systems.

We found that change was the norm in both the Army and the Air Force. Each of the development and procurement programs we studied experienced at least one change in funding. And for the great majority, actual funding differed from planned funding in every year.

The net result of the more recent funding changes was generally higher actual expenditures. Overall, the funding for Army development programs increased by 20 percent. Similarly, the Air Force added 14 percent to its planned development program spending. In procurement programs, the Air Force added nearly $5.3 billion, or 26 percent to its planned funding. In contrast, the Army cut about $636 million, or 4 percent of the procurement spending planned for FY 2000 through FY 2004.

Implications

This study was motivated by concerns that decisions made within the Army during program and budget reviews were having unanticipated and unintended effects on program execution. Our research examined three types of evidence for this issue and found little indication that this was the case. Our analysis of the relationship between funding instability and adverse program results (such as cost growth and schedule slippage) found only a single reliable association: Procurement program funding instability is associated with schedule slippage. Our three case studies suggest that external events—i.e., events beyond the control of Army leaders—were the most important sources of funding instability. Finally, an analysis of Army program funding from FY 2000 through FY 2004 showed that funding instability had increased in development programs but decreased in procurement programs. Additionally, a comparison of recent Army and Air Force funding instability found generally similar patterns in the two services.

As a whole, the evidence shows that funding deviates from plans for numerous reasons. Many programs are affected by major external events, such as geopolitical changes and reductions in defense spending. Some programs are affected by ambitious goals that cannot be achieved with available funds. Most funding instability arises not from events inside the Army, but from root causes that lie outside the Army.

Nor should one equate high funding instability with mismanagement. Large funding changes often are made for valid reasons. Changes in R&D funding may reflect an application of funds to solve unexpected technical problems; procurement changes typically arise from changes in acquisition quantities. Regardless of why they are made, changes are necessarily reflected in program funding data.

In our analyses, it was sometimes difficult to separate cause from effect. In the case of Javelin, the end of the Cold War led to reductions in Army force levels that, in turn, led to reductions in the number of Javelin missiles required. The outcome was a unit cost increase greater than 15 percent—an increase large enough to be considered a "Nunn-McCurdy breach," which must be reported to Congress. Here, the evidence of cause and effect is clear and direct. In the FAAD C2I case, the ambitious technical goals of the original program were not met, development was stretched out, and procurement was delayed. Here, the root cause was the ambitious Army decisions made in 1986 and the next few years. In these two cases, funding instability was the result of program problems, not the cause of them. Based on this evidence, we can say that funding instability, per se, is not an important cause of Army program managers' problems.

Acknowledgments

This research on the causes and effects of funding instability in U.S. Army acquisition programs could not have been accomplished without the documentation, insights, suggestions, and cautions offered by participants in the Army programs we studied and a number of our RAND colleagues. LTC Calvin Gramlich, Office of the Assistant Secretary of the Army (Acquisition, Logistics, and Technology), played a pivotal role as project officer, facilitating the research in every phase of the study.

Our review of the Javelin program benefited from discussions with LTC Philip Carey, Javelin project manager in the Close Combat Weapon Systems Program Management Office, and with Lloyd Olson, cost and budget analyst for the Javelin program. We also greatly appreciate the insights provided by two former Javelin program managers, COL (Ret.) Michael A Roddy, Aerojet Corporation, and COL John Weinzettle, Office of the Program Executive Officer, Tactical Missiles. Discussion with William Y. Bishop, formerly chief engineer of the Javelin program, also provided important insights into the technical problems that the program encountered.

Many members of the Forward Area Air Defense Command, Control, and Intelligence (FAAD C2I) Program Office helped in providing data for this project. Yolanda E. Hodge, FAAD C2I program manager, generously shared program resources and responded to our requests for program information. Jimmy D. Preston, chief of program operations, and his staff were meticulous in their efforts to ensure that we had the right information and clearly understood the data provided.

Keith Mack and Joe M. Hardin gave independent confirmation of pivotal details. We thank each of them.

LTC Anthony Pelczynski, CH-47F project manager, and members of his staff, including Janet Fletcher and John Mull, provided much useful information about the issues that arose with the program. We appreciate their assistance, as well as that of COL Tim Crosby, Cargo Helicopter program manager within the Office of the Program Executive Officer, Aviation, and former CH-47F project manager. We also wish to thank Brian Craddock, Department of the Army system coordinator for CH-47F, for helping with our research.

Several RAND colleagues made important contributions to the conduct and reporting of this work. Lauri Zeman guided this work from the start. Fran Lussier worked to get the project defined and the initial work focused. Rob Leonard shared the results of his work comparing Army acquisition performance with that of other services. Mary Tyszklewicz, Andrea Mejia, and Christine San helped gather, organize, and prepare our quantitative work. Jerry Sollinger initiated the drafting of this report and helped hone the finished product. Natalie Ziegler provided sure and responsive administrative support to the project.

Greg Hildebrandt and Neil Singer provided careful professional critiques of this report. Their comments stimulated additional analysis that did much to improve the final product.

Abbreviations

AAWS-M	Advanced Antitank Weapon System–Medium
AFM	airframe
ASARC	Army Systems Acquisition Review Council
C2	command and control
C2I	command, control, and intelligence
CBO	Congressional Budget Office
CLU	command launch unit
CONUS	continental United States
DoD	Department of Defense
DSARC	Defense Systems Acquisition Review Council
ECCM	electronic counter-countermeasure
ECM	electronic countermeasure
EMD	engineering and manufacturing development
FPA	focal plane array
FRP	full rate production
FSD	full-scale development
FUE	first unit equipped
GAO	General Accounting Office (now called Government Accountability Office)
GBS	ground-based sensor
GFE	government furnished equipment
GPS	global positioning system
IFF	identification friend or foe
IOT&E	independent operational test and evaluation
LRIP	low-rate initial production

MTS	masked target sensor
NATO	North Atlantic Treaty Organization
NCTR	non-cooperative target recognition
OPA	Other Procurement, Army
OSD	Office of the Secretary of Defense
P3I	Preplanned Product Improvement
PEO	program executive officer
PHID	positive hostile identification
PQT	product qualification test
RDT&E	research, development, test, and evaluation
ROC	required operational capabilities
SAR	*Selected Acquisition Report*
SDD	system development and demonstration
SHORAD	Short Range Air Defense
SOCOM	Special Operations Command
SOF	Special Operations forces
TBM	theater ballistic missile
UH	utility helicopter

NOTE: Descriptions of the Army and Air Force weapon systems included in the analysis are in Appendixes A and B, respectively.

CHAPTER ONE

Introduction

Background and Purpose

Each year, senior-level decisionmakers change the funding allocation to individual weapon system programs. Changes occur for many reasons: revisions to priorities and requirements, the emergence of new opportunities, technical difficulties, engineering changes, contractor management problems, acquisition budget reductions, etc. Most substantial changes are implemented during annual programming and budgeting cycles. Recommendations to make changes are reviewed by Army leaders and may also be submitted for review and approval by the Office of the Secretary of Defense (OSD). Decisions are based on the best information available at the time; mistakes can be made, however.

Some Army officials are concerned that changes in the funding allocated to individual weapon system programs may not be accompanied by adequate attention to how they affect the management of the programs. Professional acquisition officials know how funding instability affects program management in general, but those responsible for funding changes may not have this knowledge . Moreover, circumstances in individual programs can differ so much that even knowledgeable senior officials may not fully appreciate how funding changes are likely to affect the management of any particular program.

This project sought empirically based information from the experience of recent Army weapon system programs that could be used to clarify how changes in their funding have affected them. We present this information in a form that will help senior officials who are not extensively experienced in weapon system acquisition better appreciate the effects of funding instability.

Approach

Two research options exist for establishing links between funding instability and adverse effects: statistical analysis and case studies. Both seek empirical information to identify defensible and predictable links between funding instability in programs and the outcomes of those programs using analytic methods. Statistical analysis requires comparable data on a sufficient number of programs to identify significant trends and correlations. Case studies dig more deeply into the funding and operations of specific programs to identify specific program management reactions to funding changes. The analysis reported here involves statistical analysis of 18 Army programs and case studies of three of those programs that experienced high funding instability.

Although "instability" and program management have attracted a great deal of attention in the past, our review of prior research on this topic revealed no standard definition of funding instability. So we developed a new metric and used it to measure and compare funding stability in different programs. In effect, the metric summarizes differences between actual funding profiles and those estimated when the decision to begin system development is made (Milestone B). Unless otherwise noted, all cost and budget data used in this monograph are in terms of Budget Authority reported in fiscal year (FY) 2004 dollars.

The quantitative analysis also needed clear definitions of "adverse effects" to determine how large an effect, positive or negative, could be associated with funding instability. Standard metrics for adverse effects are available; we used three of those associated with cost and schedule growth. We focused on cost growth and schedule growth because they have been the standard measures of program management success used in the past. However, cost, schedule, and system performance are necessarily interconnected; it is often the case that a change in one of these factors affects the others.

From the 18 systems used for the quantitative analyses, we chose three for the case studies. The work in these studies relies on formal program reports, open source literature, and interviews with present and former program management personnel. The small number of cases implies that the findings cannot fully identify all reactions of program managers to funding instability or determine likely cause-and-effect relationships. The small number of cases also makes generalization of the results risky.

Programs Included in Quantitative Analysis

Table 1.1 provides summary data on the 18 Army programs we reviewed. The programs listed are for vehicle, helicopter, missile, munitions, and electronic systems, and each one meets the following criteria:

- Is "recent"—that is, has been active since 1998.
- Has at least five years of experience since Milestone B or its equivalent.[1]
- Has enough planning and actual data to calculate relevant metrics.

The data for this analysis were drawn from *Selected Acquisition Reports* (SARs). The Department of Defense (DoD) uses SARs to report to Congress summary information on the history of and its current plans for cost, schedule, and system performance on large acquisition programs. Since SARs are prominent reports on prominent programs, they are carefully prepared and reviewed in DoD prior to submission. They must be provided annually for all programs with total estimated

[1] We use the current acquisition terminology set forth in DoD Directive 5000.1, May 12, 2003, throughout this document. In accordance with this terminology, Milestone B, as of 2000, represents the start of the system development and demonstration (SDD) phase of the DoD system acquisition process. It is defined somewhat differently than the Milestone II that was used before 2000, which was considered to be the start of the engineering and manufacturing development (EMD) phase. For our analyses, we treated these two milestones as comparable, so Milestone B is used throughout this monograph to mean Milestone B or an earlier equivalent. Similarly, Milestone C, the current designation for the start of the production phase, is used to mean Milestone C or its earlier equivalent, Milestone IIIA, which was the authorization to start low-rate initial production.

Table 1.1
The 18 Army Programs Selected for Analysis

Program	Type	Milestone B Date	Value at Milestone B (FY04 $B)	SDD Duration (months)
Family of Medium Tactical Vehicles (FMTV)	Vehicle	Oct 1998	8.57	27
Bradley Fighting Vehicle System Upgrade	Vehicle	Dec 1993	3.10	21
Black Hawk Utility Helicopter (UH-60A/L)	Helicopter	Aug 1972	7.13	56
Longbow Apache Airframe (AFM)	Helicopter	Dec 1990	5.81	52
Chinook Improved Cargo Helicopter (CH-47F)	Helicopter	May 1998	2.76	43
Javelin	Missile	Jun 1989	4.14	36
Guided Multiple Launch Rocket System (GMLRS)	Missile	Nov 1998	2.22	35
Longbow Hellfire	Missile	Dec 1990	2.01	52
Brilliant Anti-Armor Technology Preplanned Product Improvement (BAT P3I)	Munition	Jun 1991	2.27	41
Sense and Destroy Armor Submunition (SADARM 155)	Munition	Mar 1988	2.08	47
Advanced Threat Infrared Countermeasure/Common Missile Warning System (ATIRCM/CMWS)	Electronic	Sep 1995	2.92	55
Forward Area Air Defense Command, Control, and Intelligence (FAAD C2I)	Electronic	Sep 1986	1.94	20
Longbow Apache Fire Control Radar (FCR)	Electronic	Dec 1990	1.64	52
Joint Surveillance Target Attack Radar System Ground Station Module (JSTARS GSM)	Electronic	Aug 1989	1.51	40
Secure Mobile Anti-Jam Reliable Tactical Terminal (SMART-T)	Electronic	Nov 1992	0.91	40
Joint Surveillance Target Attack Radar System Common Ground System (JSTARS CGS)	Electronic	Aug 1993	0.86	84[a]
AN/TYQ-45 Maneuver Control System (MCS)	Electronic	Oct 1989	0.61	30
Combat Service Support Control Systems (CSSCS)	Electronic	Feb 1991	0.30	30

SOURCE: RAND calculations using *Selected Acquisition Report* data.

[a] Actual duration; planned duration is not available.

costs greater than $365 million for development or $2.18 billion in total cost (both in FY 2000 dollars). SARs are the most widely used central database for analysis of program cost and schedule.[2]

Unless otherwise noted, all data in Table 1.1 reflect plans at Milestone B. The third column provides the date of Milestone B, which occurred after 1985 for 17 of the 18 programs. That was the year that DoD and Army budgets peaked, so the data for the individual programs cover years of continuing fiscal stringency. The fourth column includes the expected value of the program at Milestone B. The 18 programs vary in size from the largest, at $8.57 billion, to the smallest, at $300 million; the average planned size of the selected programs is substantial, at $2.74 billion. The last column, with one annotated exception, shows the expected duration of the SDD phase—that is, the time in months between Milestones B and C (as planned at Milestone B). Planned system development duration also varies significantly, from 20 months to 56; the average planned duration is just less than 40 months. Clearly, there is considerable variation in the planned size and duration of major Army programs.

We were unable to include several prominent programs in our quantitative analysis. In some cases (the M1A2 Abrams tank, for example), the SAR entry was missing important data. In other cases (Stryker, for example), the program was too recent to have accumulated enough experience for us to examine.

Metrics for Program Changes

We used simple measures of cost growth and schedule growth drawn from the literature as the key adverse effects to be considered. But, as the findings of our quantitative analysis show, the outcomes of cost and schedule growth were not always adverse. Two programs beat their scheduled goals, and three delivered their products at a cost lower than was expected at Milestone B.

A development cost growth factor was used to measure increases in development cost. This factor uses costs that have been adjusted

---

[2] For additional information on *Selected Acquisition Reports*, see Office of the Under Secretary of Defense (Acquisition, Technology, and Logistics), 2003.

to reflect the same price level (that is, constant dollars) to wash out any effects of inflation. It divides the cost of research, development, test, and evaluation (RDT&E) at a given date by the cost of RDT&E anticipated at Milestone B. We used Milestone B as a baseline for the same reason that most past analyses have used it: Generally, program data do not become consistent and reliable until about the time of Milestone B.

A procurement cost growth factor was used to measure unexpected increases in procurement cost. This factor divides procurement cost at a given date by the procurement cost estimated at Milestone B. Though procurement cost growth, like development cost growth, is measured in constant dollars, its estimation is more complex, because changes in procurement quantities after Milestone B must be taken into account. To adjust for quantity changes, we adopted a standard approach.[3] Data from a program's most recent SAR report were used to estimate a cumulative average cost learning curve, which was then used to estimate the cost of the Milestone B baseline quantity planned for procurement. The procurement cost growth factor reported here for *x* years after Milestone B is the total program procurement cost implied by dividing the learning curve in year *x* at the level of production assumed in the Milestone B report by the total program procurement cost Milestone B.[4]

We used schedule slippage between Milestones B and C as the measure of schedule growth (the acquisition literature normally measures slippage over a longer period). To focus on recent Army programs, we used a common end point for the schedule measure to keep the sample we examined as large as possible. The use of a milestone later than Milestone C would have undercut our quantitative analysis by limiting the number of programs it included.

---

[3] Hough (1992) explains alternative approaches to normalizing for quantity changes. The Institute for Defense Analyses has used the same measure in its studies. The SAR process uses a different measure but provides the data required to execute the RAND method.

[4] In this formulation, a change in program scope can result in development cost changes as well as procurement changes. The lack of a metric for program scope means that we cannot adjust for scope changes.

Organization of This Monograph

Chapter Two begins with a review of the literature on acquisition program performance and management. It then continues with definitions and estimates of funding instability and addresses the association of funding instability with adverse program outcomes. Chapter Three describes the results of our case studies of three relatively recent Army acquisition programs: Javelin, FAAD C2I, and CH-47F. The case studies look into the nature of the causes and effects of funding instability and seek to determine whether funding changes initiated by Army leaders are an important source of program management problems. Since these analyses used data for Army programs active in the 1990s and thus produced results that do not reflect more-recent experience, Chapter Four examines funding instability in Army and Air Force programs since 2000. Chapter Five presents a summary of our key findings. Two appendices provide descriptions of the programs included in our reviews: Appendix A for the Army programs, and Appendix B for the Air Force programs.

CHAPTER TWO

Quantitative Analysis of Army Programs

This chapter describes the quantitative approach we used to analyze the management of DoD weapon system acquisition programs. It begins with a review of the literature about cost and schedule growth in acquisition programs and then goes on to describe how funding instability might affect program cost and schedule. Finally, it reviews recent Army experience and looks for associations between funding instability and cost growth or schedule slippage.

What the Acquisition Literature Says

Although the body of literature on the performance and problems of weapon system acquisition programs is substantial, going back at least four decades, it contains little that directly addresses funding instability. Many of these studies address "adverse effects" relevant to weapon system programs; others develop and advocate various acquisition reforms. The adverse effects discussed take several forms, predominantly cost growth, schedule slippage, and less-than-expected weapon system performance, all of which can occur for many reasons. This section summarizes what recent empirical studies offer as insights about schedule and cost growth in DoD acquisition programs.[1]

---

[1] The literature on weapon system performance failures is relatively thin. While cost and schedule are readily measurable dimensions of program performance, the performance of the weapon system produced is multidimensional and hard to characterize in a single summary

Schedule Growth

The literature indicates that schedule growth can stem from three sources: changes in program guidance, funding, and system performance requirements.

Changes in program guidance from outside agencies are common; they affect the schedules of almost all programs. "External" guidance can come from Army leadership, OSD, Congress, and/or the White House. Such guidance usually lies beyond the program's control. Drezner and Smith (1990) indicate that instability caused by external direction delayed schedules for ten DoD programs started during the 1970s and 1980s by an average of over six months.

Funding changes are also common, and they can have large effects on schedules for longer programs. Drezner and Smith (1990) found that funding instability delayed the actual schedule relative to the plan by an average of 4.6 months. A 1987 study by the Congressional Budget Office (CBO) found that the level of funding actually available during the procurement phase drove the production rate and, through this effect, the schedule slippage later in a program.[2] The longer planned production continued, the greater the potential for such delays.

Changes in system performance requirements—either the system specifications or the quantities to be produced—can also have large effects in longer programs. The longer a program persists, the greater the potential for new demands to be placed on the program, altering the initial plan and delaying the initial schedule.

Cost Growth

The literature we reviewed told us less about the sources of cost growth. In general, program stability, defined using any of the perspectives above, tends to limit cost growth.

Technical complexity is the other major factor related to cost growth: The more complex the program, the more cost growth should be expected. Cost growth may be associated with ambitious system

---

measure. Moreover, many key performance parameters are unique to the particular system, making comparison of performance parameters across systems a daunting task.

[2] Also see Congressional Budget Office, 1982.

performance goals. Air Force Systems Command research (U.S. Air Force Systems Command, 1982) suggests that cost growth was more affected by technical complexity than by program stability in programs under way before 1970 and that program stability was the more important of the two during the 1970s. The reasons for this reversal are not clear. Nor is it clear whether this reversal persists.

Cost growth studies on weapon systems span decades, as cost growth has been a concern for many years. Some studies focus on cost growth in a particular type of system, such as aircraft, while others include many types of systems. Examinations of cost growth for weapon systems are also often included in larger-context studies on cost estimating and acquisition reform. The literature addresses a multitude of aspects of cost growth, including magnitude and causes.

A General Accounting Office (GAO) study (GAO, 1994) and a RAND study (Drezner et al., 1993) reported that weapon system cost growth had commonly been in the range of 20 to 40 percent since the 1960s. The same sources showed that many programs exceeded the 40 percent range by a large margin. A RAND study that evaluated cost growth on a collection of defense systems showed development cost growth ranging from 0.4 to 85 percent and procurement cost growth ranging from 18 to 100 (Birkler et al., 2001). Calculations show that for the same systems, the mean development cost growth was 26 percent and the mean procurement cost growth was 60 percent. These results confirm that there is still great variation in cost growth but that the range has been fairly stable.

Stark (1973) used a dollar-per-pound measure to argue that the cost growth for fighter aircraft was six times higher than that for commercial aircraft from 1940 to 1970. He argued that the trend toward keeping military aircraft in the inventory longer led to a tendency to push the state of the art, which could double the procurement cost. In addition, Stark cited requirement changes and production schedule stretch-outs as other major causes for cost growth.

Another GAO report (GAO, 1992) included technical risk, funding inconsistency with goals, and optimistic cost and schedule estimating as causes of cost growth. In 1998, the Secretary of Defense listed similar causes of cost growth in his 1998 annual report to Congress, citing technical risk, schedule slips, optimistic cost estimating, and volatile funding profiles as primary causes (Cohen, 1998).

Phillips (2004) examined cost growth for aircraft weapon systems, comparing the average annual cost growth on 11 programs from a pre-reform period (1991–1996) to the average annual cost growth on seven programs from a post-reform period (1997–2001). He found that there was no significant difference in this cost growth between the two periods and concluded that DoD reforms during 1991 to 2001 do not appear to have significantly stemmed cost growth.

How Funding Instability Might Affect Cost and Schedule

The literature also suggests "how" funding instability could affect program cost and schedule growth. "How" can be a subtle concept, however, and is best inferred from the many case studies available on the management of individual DoD weapon system programs. As a whole, this literature points to three "layers" of effects.

The first-layer, or most obvious, effect is what occurs when changes in funding require the actual quantity procured to depart from the planned quantity: schedule and unit cost change immediately. A reduction in quantity allows the program to be completed more rapidly and an increase has the opposite effect.[3] A reduction in quantity also increases the importance of development costs relative to production costs and thereby drives up the unit cost of a program. These effects are well understood and fairly easy to predict and quantify, at least to a first order of approximation.

The second-layer effect, which is somewhat subtler, is that unanticipated funding changes of any kind tend to reduce the value of past investments in a program. Resource costs and time incurred to implement past plans have been wasted to some extent; they have been invested in things no longer valued. So the effective investment in the program has fallen. To get back to par and meet the new requirements imposed by the unanticipated change, the program must incur addi-

[3] Although a reduction in quantity may allow a program to be completed more rapidly, procurement stretch-outs to reduce spending rates often thwart the achievement of earlier completion dates.

tional and unanticipated costs and delays in the future. Put another way, cost and schedule both grow. And because they are to some extent substitutes for each other in most programs, program managers can affect how much each one grows. Thus, if staying on schedule is important, the program can usually spend more to make that happen; if cost containment is important, the schedule may shorten.

The third-layer effect is subtler still. For many reasons, a schedule slip can be expected to raise costs. For example, if the fixed costs of a management team, of program-specific facilities and equipment, or of a fixed work force must be sustained for a longer time, total costs obviously will rise. If many activities had been optimized in a jointly dependent schedule, a schedule slip forces changes in that coordinated solution. Such changes typically impose costs to rearrange resources and to ramp operating activities up and down. These costs occur on top of the costs of fixed assets suspended in limbo during the adjustment. For example, subcontracts often have explicit terms or equitable adjustment clauses that allow subcontractors to pass on costs incurred to reorient and reschedule inputs to a DoD program. In most weapon system programs, these costs pass directly to the government.

How these layers of effects occur in any particular setting depends on the policy context in which DoD and the Army pursue a program. One should expect these effects to change over time as the policy context changes. For example, if DoD programs place a high priority on the Cost As an Independent Variable initiative, one would expect them to tolerate greater slippage in weapon system performance in order to avoid cost growth. This suggests that cost growth would be less likely to occur than it has been in the past relative to schedule growth. Put another way, just because these effects have been seen in the past, one should not expect to see them persist in similar ways in current Army programs.

Past empirical studies have often asked whether cost growth and schedule growth tend to occur together or whether one falls when the other rises.[4] The discussion above indicates that within a program, the relationship between the two depends, in fairly predictable ways, on

---

4 Jaynes (1999), for example, ultimately focuses on this question.

the source of instability that caused the program adjustment and on the circumstances inside the program when the adjustment occurred. But one should expect the answer to differ across programs and should not be surprised that past studies have detected no strong patterns in the relationship between cost growth and schedule growth across DoD programs.

How Army Program Management Compares with That of Other Services

Decisionmakers and program managers often address the fundamental trade-offs among program schedules, costs, and system performance parameters. Schedules and cost goals are more readily measured than system performance, which depends on a complex of as many as 15 or 20 specified parameters.

This section compares the Army's performance on development cost, procurement cost, and achievement of schedule goals with that of other services and agencies managing their major DoD programs. The database we used for this comparative analysis came from another RAND project, which drew from SAR records for programs active in the 1990s and early 2000s.[5] A handful of those programs were begun in the 1970s, although most were initiated later; and the sample of Army programs included most of the programs we used (see Chapter One, Table 1.1). The other RAND project's sample and time period are not identical with those we used in our analysis of financial instability. Data for Air Force and Navy Department programs were taken from the same source for programs active during the same period.

Development Cost Growth

Figure 2.1 shows the development cost growth factor for different periods of time following Milestone B for both Army and non-Army pro-

---

5 This section draws importantly on the unpublished work of our RAND colleague Rob Leonard.

Figure 2.1
Army Development Cost Growth Compared with That of Other Services

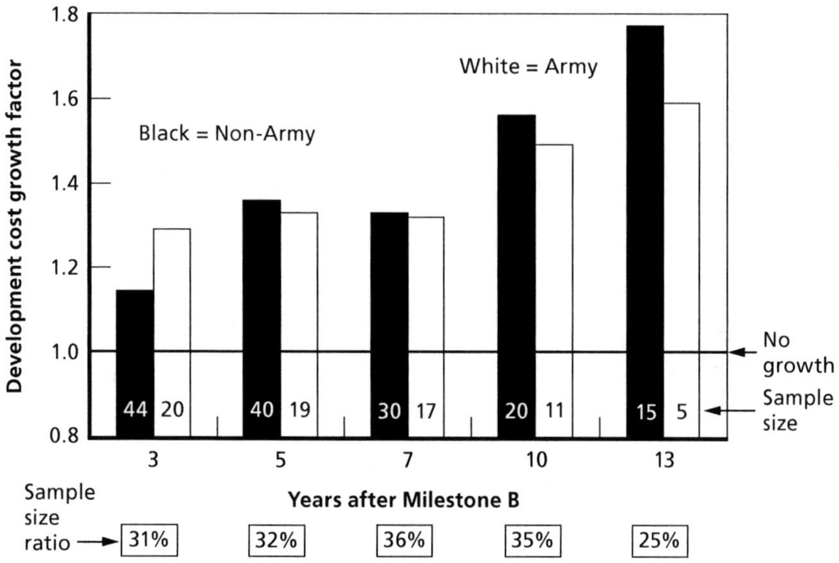

SOURCE: Unpublished RAND research.
RAND *MG447-2.1*

grams.[6] If no cost growth occurs, the expected cost remains constant, and the cost growth factor in the chart equals one.

For each period of time, the chart shows an Army and a non-Army (other service) cost growth factor, corrected for inflation. The numbers in each bar denote the number of programs included in the average. The sample size ratio at the bottom of the chart shows the Army's share of the total number of programs in each comparison.

Figure 2.1 shows two major results. First, it shows that development cost grows substantially beyond Milestone B in the Army and elsewhere in DoD, and generally continues to grow with each passing year after Milestone B. In other words, the further along the program, the greater the cost growth. In effect, additional events after Milestone

[6] Development cost growth is calculated from Milestone B for periods of three, five, seven, ten, and 13 years (shown from left to right in Figure 2.1). Procurement cost growth (shown in Figure 2.2) was also calculated from the Milestone B baseline for the same periods. See Note 1 in Chapter One for an explanation of milestone designations.

B occur with each passing year, and these events are systematically more likely to raise than to lower development costs each year. This pattern has persisted in DoD as a whole since at least 1960, despite repeated efforts to improve cost estimation and to control costs.[7]

Second, it shows that development cost grows somewhat faster in Army programs than in other DoD programs in the first years following Milestone B, at about the same rate five to seven years out, and somewhat more slowly after seven years. Since the development phase tends to be complete within seven years, the average Army performance and DoD performance are quite similar for the majority of major programs.

What Figure 2.1 does not show is another major pattern in DoD as a whole—i.e., that the actual cost growth factor across programs within each service varies dramatically from the averages reported here. This variation is far larger than any difference between the Army and the rest of DoD for any horizon.[8]

Procurement Cost Growth

Figure 2.2 shows results analogous to those in Figure 2.1. In this case, it shows the procurement cost growth factor, corrected for inflation and normalized for changes in quantity procured, for different periods following Milestone B.

There are two important results here as well. First, Figure 2.2 shows that procurement cost grows substantially beyond Milestone B in the Army and elsewhere in DoD, and continues to grow with each passing year after Milestone B. The pattern and the explanation for procurement cost growth are much the same as those for development cost growth. And, as with development cost growth, this pattern

---

7 A similar pattern occurs in complex commercial development programs (for evidence on cost growth in first-of-a-kind pioneer plants in the chemical and petroleum industries, see Merrow, 1989, and Merrow, Phillips, and Myers, 1981). That is, the pattern of growth here is not unique to DoD. It is, at least to some extent, inherent in the development of complex technologies.

8 More details on this variation are in Drezner et al., 1993, and in an unpublished document on cost uncertainty by Wong, Drezner, and Hess.

Figure 2.2
Army Procurement Cost Growth Compared with That of Other Services

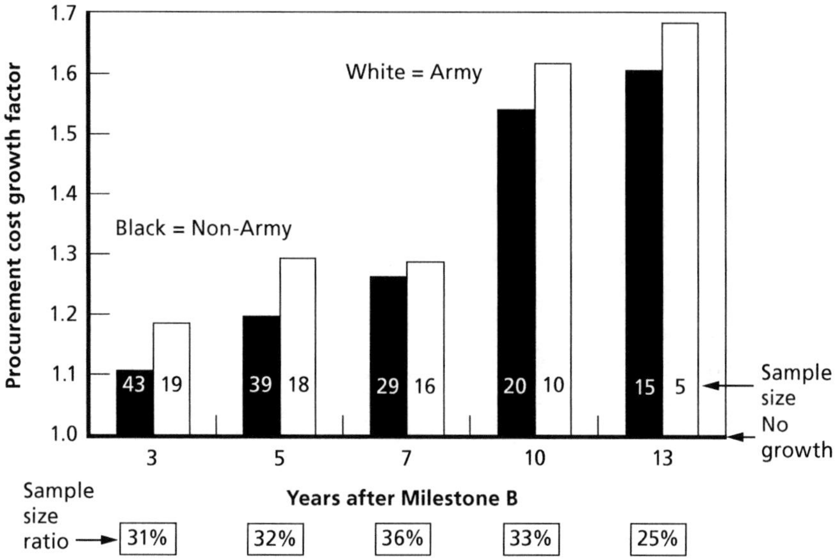

SOURCE: Unpublished RAND research.
RAND *MG447-2.2*

has persisted in DoD since 1960, despite repeated generations of management reforms aimed at reducing cost growth.

Second, procurement cost growth is a bit higher in the Army than in the rest of DoD at every year following Milestone B. But in every year, the differences between the averages are much smaller than the variations of cost growth among individual programs in the Army and elsewhere in DoD. Moreover, past analyses (Tyson et al., 1989, 1992) suggest that if we had controlled for specific attributes of the Army and non-Army programs, such as dollar size and complexity, the differences would be somewhat smaller.[9]

---

[9] More details on this variation are in Drezner et al., 1993, and in an unpublished document on cost uncertainty by Wong, Drezner, and Hess.

Program Schedule Changes

Figure 2.3 uses the length of time between Milestones B and C (in months) to measure program length.[10] It shows the average actual Milestone C dates (second set of bars), the average Milestone C dates planned at Milestone B (first set), and, as in Figures 2.1 and 2.2, the number of Army and other service programs included in the averages.

As can be seen, the planned schedules for getting from Milestone B to Milestone C are similar for the Army and other service programs. However, as the last pair of bars shows, the Army schedules slipped 21 months, or 50 percent, more than did the non-Army schedules in the sample.

As with cost growth, schedule slippage is large in the Army and elsewhere in DoD. This finding is comparable with the findings of other recent studies. Variation in program-specific schedule slippage is very high around these averages. The difference in Army and non-Army

Figure 2.3
Army Schedule Slippage Between Milestones B and C Compared with That of Other Services

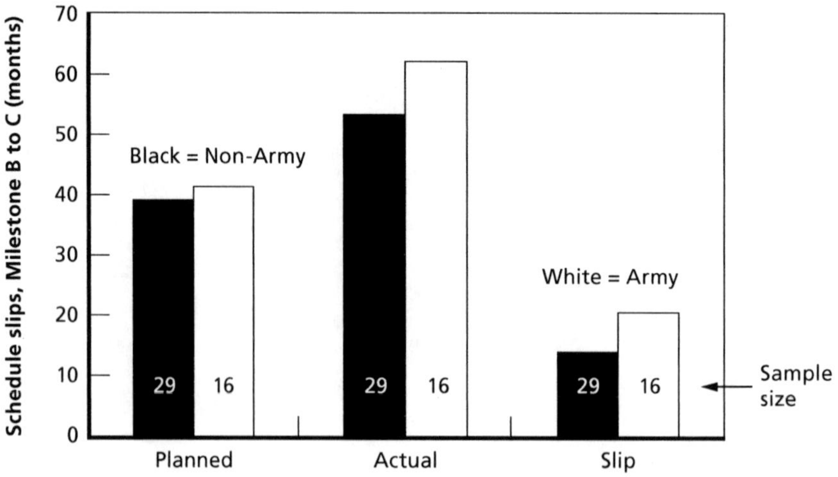

SOURCE: Unpublished RAND research.
RAND *MG447-2.3*

---

[10] See Note 1 in Chapter One for a discussion of milestone designations and definitions.

schedule slippages is significant but should be kept in perspective. It is moderate relative to the average level of slippage and small relative to the variation in slippage throughout DoD.

Funding Instability, Cost Growth, and Schedule Growth in Army Programs

This section examines data from the 18 Army programs identified to seek associations between funding instability and cost and schedule growth. Careful analysis of links between funding instability and adverse effects benefits from clearly defined measures. Although analysts have discussed instability in DoD programs for a long time, they have not developed funding instability measures that could help address the policy problem. Our analysis offers a new measure that captures in one number the difference between planned and actual funding.

We define development funding instability as the sum of the absolute value of the differences between planned funding and actual funding over the period.[11] Planned funding levels for the first five years of development are taken from the initial SAR containing the developmental acquisition program baseline. Actual funding levels for the same period are taken from a recent SAR. To compute the funding instability score for a program, the first step is to convert both sets of data to FY 2004 dollars. Second, the differences between the actual and planned funding are calculated for each year. Third, the absolute values of the differences are summed for the five years and divided by the funding planned for the same period.[12] If the base year is too recent to record five years of actual differences, we complete the five

---

[11] The development program for a system is intended to accomplish the engineering design and testing required to transform a promising concept into an effective and producible system design that meets a defined set of performance parameters. Thus, development logically precedes procurement. The initiation of system design and development is typically preceded by concept and program definition efforts that may include development of a prototype. It is hard to tell whether the systems approved to move forward at Milestone B are at the same stage of development. But success in development is likely the first priority for program managers just after passing Milestone B reviews.

[12] Values for financial instability can range upward from zero (no difference between actual and planned funding) to any positive number.

year series for "actual funding" with planned funding from the most recent SAR.

Figure 2.4 illustrates the instability calculation for the Javelin missile development funding. In the case of Javelin, the measure of financial instability is 86.2, a percentage of the research funding planned for the Javelin at Milestone B.

The process for calculating procurement funding instability is identical except that data cover the first five years of planned procurement. Procurement often begins with "low rate" buys to allow the manufacturing process to be refined and the service users to test the new system before the service commits to more expensive, large buys. Thus, the procurement period is generally later in time than the first five years following Milestone B.

The funding instability measure is simple and readily calculated and relevant. It uses funding planned for the program at the Milestone B review as the baseline and compares it with the actual funding provided later to execute the program. But it does not necessarily reflect funding instability as the program manager would experience it. The measure shows the net results of all changes to the program, not necessarily the frequency of changes. Three other aspects of this measure that need to be discussed here are the five-year period, absolute values, and the aggregated total.

Figure 2.4
Example of Funding Instability Calculation: Javelin RDT&E

Javelin calculation (in FY 2004 dollars):

	1990	1991	1992	1993	1994	Total
Planned RDT&E	199.8	142.8	43.4	4.8	1.7	392.5
Actual RDT&E	174.0	97.8	146.6	116.7	54.3	589.4
Difference	–25.8	–45.0	+103.2	+111.9	+ 52.6	+196.9
Absolute value	25.8	45.0	103.2	111.9	52.6	338.5

Javelin development funding instability score:

Absolute value/Planned research = $338.5/$392.5 = 86.2

RAND *MG447-2.4*

For our analysis, we began by examining SARs for over 30 Army programs. That set was reduced to 18 programs largely because there were too few data on actual procurement spending for five years. While a period longer than five years would have provided more data for each individual program, the number of programs included would have been further reduced. Our choice of five years is thus a compromise between depth of information on individual programs and breadth of analysis across programs.

The use of absolute values means that increases and decreases to program funding are treated as equally important. Though project managers are likely to view increases far more favorably than decreases, unplanned increases in funding after Milestone B do force managerial adjustments, just as decreases do. Moreover, increases and decreases may be related, as in cases in which program schedule slippages lead to cuts from early spending and increases in later spending.

The aggregation into a five-year sum conceals the patterns of net changes within individual programs that could provide additional insights. A review of these patterns showed that net changes in program funding are the norm (see Table 2.1, top panel). For the 18 programs we selected, there were 180 yearly observations (five each for research and for procurement). One Army research program's funding went unchanged for two years after Milestone B; the other 178 observations all showed differences between funding approved at Milestone B and funding actually provided. Notably, 17 of the 18 RDT&E programs had their funding changed in the year following the intensive Milestone B review conducted within the Army and OSD. Increased funding was slightly more common than reductions in research program funding. In procurement programs, decreases from planned funding were notably greater than increases.

But the differences between planned and actual funding were not all of equal magnitude. In Table 2.1 (bottom panel), an arbitrary standard of 10 percent is used to highlight changes between planned and actual funding in each year that could be called significant. In the RDT&E programs, a little more than one-half of the changes were greater than 10 percent, and they were equally divided between increases and decreases. For procurement programs, 90 percent of the

Table 2.1
Year-by-Year Differences Between Planned and Actual Program Funding

	RDT&E Programs					Procurement Programs				
	Year After Milestone B					Year of Planned Production				
Differences	1	2	3	4	5	1	2	3	4	5
All										
Reductions	8	10	9	8	6	13	13	10	11	11
Increases	9	7	9	10	12	5	5	8	7	7
Total	17	17	18	18	18	18	18	18	18	18
Greater than 10%										
Reductions	4	4	2	8	5	11	13	10	11	11
Increases	2	2	5	3	11	2	4	6	7	6
Total	6	6	7	11	16	13	17	16	18	17

changes were greater than 10 percent and nearly 70 percent of the total were reductions from planned funding. The first year of planned procurement was later zeroed out for eight of the 18 Army programs. For five of the eight, funding was also eliminated in the second year of planned procurement.

This review of the patterns of change within the 18 Army programs shows that changes between planned and actual funding are the norm. RDT&E programs tend to get more increases than reductions, and the reverse is true for procurement programs. It appears that ambitious research programs' funding requirements tend to be underestimated; as a consequence, procurement slipped.[13]

Given these definitions and conditions, we used three measures of instability to capture different elements of funding in our analysis. The development measure applies the formula illustrated in Figure 2.4 to the first five years of RDT&E funding for the program following the Milestone B decision. The procurement measure uses the first year with planned procurement spending, as identified in the Milestone B acquisition program baseline. The total measure combines the development and the procurement funding instability into a single measure.

---

[13] In 14 of the 18 Army programs, procurement was to begin within three years or less of the Milestone B review and decision.

When applied to individual programs, these measures for funding instability reflect financial changes, regardless of their source. The measure of funding instability can represent unexpected successes, unanticipated difficulties, or events and root causes from outside the program and the Army. If all funding is supplied on the schedule approved at Milestone B, the absolute value of the differences and the instability score will be zero. However, if development proceeds faster or more completely than anticipated, the metric for development funding instability will rise, because the originally planned development activity will not be required. If the experience is the converse, and early development efforts consistently fail, the program may be cut back or killed, in which case the measure of funding instability will increase. Again, the result in all cases is funding instability. As a final example, suppose that an earthquake were to devastate a manufacturing plant in Southern California. The program's funding would be changed in this case, too, and funding instability would be the result.

The Javelin example displayed in Figure 2.4 shows changes to development program funding in each of the five years examined. This suggests that change from planned development program funding is the norm for Javelin program managers. We found that change from planned funding was common in the three programs chosen for case studies. There are ten annual observations for each of the three case-study systems, five for RDT&E funding and five for procurement funding. Every one of these observations showed significant changes. Nine were increases; 21 were funding reductions. The smallest change in a single year was 13 percent of planned funding; the largest was a 3,150 percent increase to a relatively small planned program. In five of the 15 observations for procurement, funding was set to zero. But the three case-study systems were chosen because they had high funding instability scores.

Analyses discussed in Chapter Four examined funding instability during five years, 2000 through 2004, for both Army and Air Force development and procurement programs. Funding instability scores were found to range from as low as 1 to an extreme high of 851. The analyses included all systems for which both planned and actual totals were found in recent SARs. There were 310 annual observations, of which nearly 93 percent were changes, almost equally divided between increases and decreases in program funding. So funding changes are the norm for program managers.

Given that change is the norm, the task of this analysis is twofold. One is to see how funding instability is related to adverse or unanticipated program outcomes, i.e., to study the effects of funding instability. The second is to look into the root causes of funding instability to assess the extent to which such instability is uniquely caused by Army leaders' decisions.

We used standard measures, both drawn from earlier studies, for two types of adverse effects in our analysis. One of these is the *total cost growth factor* from Milestone B to C, adjusted for inflation and quantity change; it is used to measure cost growth. Like our predecessors, we use Milestone B because it is the earliest point at which data are likely to be available in all SAR data files. As noted above, we use Milestone C as the first clear milestone following Milestone B. The second measure is *schedule slippage*, in months, from Milestone B to C; we use this to measure schedule changes. Again, we chose these milestones to make the sample of recent Army programs we can examine as large as possible.

Changes in system performance requirements are a third type of adverse effect. Development programs generally aim to achieve improvements in operational capabilities (such as range, accuracy, detection probability, survivability, reliability, reparability, interoperability, reduced manning, and operating range) or a host of other characteristics specific to individual programs. When performance goals cannot be met, forcing developers and their sponsors to accept less capability than was expected when Milestone B decisions were made, adverse effects occur in the form of compromises. We found examples of such compromises, but there is no generally recognized method for quantifying these adverse effects in ways that enable comparisons across programs. Thus, developing a metric for performance was beyond the scope of our project.

This analysis does not constitute a complete, formal statistical analysis that controls for factors other than funding instability that might induce adverse effects. The number of observations was too small to allow such refinement. We present here exploratory statistics; we have not developed information on or metrics for other relevant factors. The associations we discuss are simple correlations between measures of instability and cost and schedule growth.

The planned funding and actual funding differ in all of the Army programs we examined. Table 2.2 shows the instability scores for development and procurement funding. As described earlier, these scores are calculated from each program's SAR data for the periods immediately or shortly after Milestone B is achieved. They compare (in

Table 2.2
Funding Instability in the 18 Selected Programs

Program	Type	Development Funding Instability Score	Procurement Funding Instability Score
FMTV	Vehicle	27.9	52.5
Bradley Upgrade	Vehicle	36.7	32.2
UH-60A/L	Helicopter	2.3	32.0
Longbow Apache AFM	Helicopter	3.5	24.3
CH-47F	Helicopter	58.6	31.1
Javelin	Missile	86.2	75.4
GMLRS	Missile	39.0	68.6
Longbow Hellfire	Missile	22.1	41.3
BAT P3I	Munition	44.1	100.0
SADARM 155	Munition	nil	88.4
ATIRCM/CMWS	Electronic	19.6	95.3
FAAD C2I	Electronic	44.7	99.9
Longbow Apache FCR	Electronic	2.0	18.2
JSTARS GSM	Electronic	32.2	33.3
SMART-T	Electronic	20.2	66.7
JSTARS CGS	Electronic	18.6	92.7
AN/TYQ-45 MCS	Electronic	6.5	62.4
CSSCS	Electronic	10.7	76.8
Average score:		26.4	60.6

SOURCE: Calculated from *Selected Acquisition Report* data.

constant dollars) actual spending over a five-year period with the spending planned for that period.

Table 2.3 shows estimates of three potential consequences of funding instability: development cost growth, procurement cost growth, and schedule slippage. Each of these measures of adverse outcomes shows variability. Development cost growth ranges from a 10 percent decrease to a 185 percent increase. For procurement cost growth, the range of estimates is even wider, from a 39 percent decrease to a 181

Table 2.3
Measures for Adverse Outcomes in the 18 Selected Programs

Program	Development Cost Growth (Ratio)	Procurement Cost Growth (Ratio)	Schedule Slippage (Months)
FMTV	2.85	1.62	9
Bradley Upgrade	1.16	1.37	17
UH-60A/L	1.02	1.14	–4
Longbow Apache AFM	1.65	2.18	8
CH-47F	1.15	2.18	12
Javelin	1.30	1.65	24
GMLRS	1.88	2.81	17
Longbow Hellfire	1.33	1.11	8
BAT P3I	1.89	1.45	55
SADARM 155	1.40	NA	NA
ATIRCM/CMWS	1.06	1.29	23
FAAD C2I	1.07	0.86	60
Longbow Apache FCR	0.98	1.37	11
JSTARS GSM	1.02	NA	NA
SMART-T	1.36	0.95	–1
JSTARS CGS	0.90	NA	NA
AN/TYQ-45 MCS	1.56	1.01	56
CSSCS	1.12	0.62	20
Average:	1.37	1.20	21

SOURCE: Calculated from *Selected Acquisition Report* data.

NOTE: NA = not available.

percent increase. And schedule slippage ranges from a gain of four months to a slip of 60 months (five years).

Statistical analyses of the data in Tables 2.2 and 2.3 show few identifiable associations. We looked for statistical associations (or correlations) between measures of funding instability (Table 2.2) and measures for cost growth and schedule growth (Table 2.3) to see whether they existed and what avenues of additional analysis might be suggested. We also recognized that simple associations might or might not indicate a cause-and-effect relationship. Identifying causal relationships remains a difficult problem in this kind of research.

In addition, we examined the association between planned program size (measured in millions of FY 2004 dollars) and funding instability. In the case of large programs, decisionmakers may cut their funding because they are thought to have greater latitude in making adjustments. Or the large programs may be protected from cuts because they have powerful constituencies or the potential to make revolutionary improvements in military capabilities. In the case of small programs, they may be kept intact for the same reasons as the large programs, or because they cannot yield enough funds—even if killed—for new initiatives. But they may also be cut because they have little visibility and few defenders. Our statistical analysis did not reveal a significant association between funding instability and planned program size as measured by planned program funding.

Figure 2.5 displays the simple association of this kind that we saw in the data: a positive, if not strong, association between total funding instability and schedule slippage.[14] Eliminating any one of the programs in the figure as an outlier—i.e., an observation in which the association is being driven by something different from the factors relevant to the other programs—would not change this basic, positive relationship. The degree of variation shown is compatible with what earlier studies found in associations of this kind. Behavior in

---

[14] This association probably flows primarily from an association between procurement funding instability and schedule slippage. There was no significant correlation of development funding instability and schedule slippage.

Figure 2.5
Relationship Between Funding Instability and Schedule Slippage

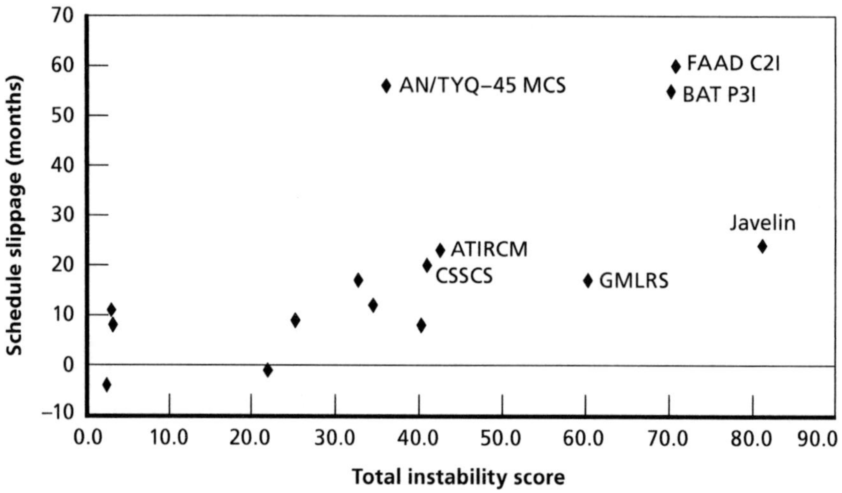

SOURCE: Calculated from *Selected Acquisition Report* data.
RAND *MG447-2.5*

individual programs always shows a great deal of variation around any average level of behavior, however defined.

Also of note in Figure 2.5 are the two programs in the upper-right corner, which have relatively high total instability and high schedule slippage. These programs offer a natural place to pursue more in-depth analysis of links between funding instability and adverse effects.

This project was motivated by a concern about the effects of high funding instability in Army programs. Our correlation analysis provided little evidence of connections between funding instability and traditional measures of program management problems. Sample sizes were relatively small, however, so this is not surprising. And correlation does not tell us about causation. Even if we had found a correlation that was high and statistically significant, a more detailed analysis would be needed to uncover the causes of Army program management problems. To determine whether funding instability is, in fact, a cause of program management problems, the connections need to be examined by other research approaches. Case studies provide an alternative approach.

What Quantitative Analysis Reveals

The literature on acquisition management shows that external guidance leads to slippage in acquisition program schedules and that technical complexity is a major factor in program cost growth. Moreover, program funding stability tends to limit cost growth. Conversely, funding instability in programs creates adverse effects through changes in quantity, changes in the productivity of existing plants and equipment, and subtle effects on management and subcontract activities. The adverse effects differ among programs, and the literature detects no strong patterns across DoD.

A comparison of adverse effects in Army and other service acquisition programs generally confirmed the lack of strong patterns. Development cost growth in all services was found to increase with time, but average cost growth for the Army and other services was much the same in most major programs. The general pattern for procurement cost growth was similar to the pattern for development cost growth, though average procurement cost growth in the Army was somewhat higher than that in other services. Schedule slippage was found to be large in the Army and the other services. In all these comparisons, variations within all services were high relative to the average.

New measures of instability were used in the analysis of data from recent Army acquisition programs. These new measures and standard measures of adverse effects revealed wide variation in all measures. Our quantitative analysis found little or no association between funding instability and adverse outcomes.

In sum, adverse effects in Army programs are generally comparable to those in other services, and funding instability in Army programs is not strongly associated with adverse outcomes. The issue of whether there is a connection between funding instability and adverse outcomes requires a more detailed analysis.

CHAPTER THREE

Case Studies of Army Programs

Additional analysis should help high-level decisionmakers understand the effects of funding instability on program management. Officials who have experience in system acquisitions are likely to already understand these effects. But because of the authoritative nature of external guidance and the fact that funding decisions are not directly motivated by acquisition concerns, many senior officials who lack acquisition experience affect the funding available for acquisition programs. This chapter seeks to develop information from three case studies to gain a better appreciation of funding instability's sources and effects.

Broadly speaking, funding instability can arise from sources within the Army or from root causes outside a program manager's area of influence.[1] Within the Army, the leadership's priorities may change (as with the ongoing transformation of the Army) and lead to funding instability. Or the complexity and ambition of approved Army programs may lead to adverse effects that in turn reveal themselves in

[1] A somewhat different perspective is contained in McNicol, 2004. McNicol distinguishes between two components of cost growth—"decisions" and "mistakes"—with the latter capturing the costs of executing the program as approved at Milestone B. The mistakes category can include optimistic cost estimates; it can also include failure to properly specify the needed characteristics of the weapon system. Mistakes tend to originate within the services, but validation by OSD at Milestone B ultimately attributes the mistakes to all participants in the Defense Acquisition Board process. McNicol estimates that the mistakes component of cost growth for the Army is 26 percent of the cost growth in engineering and manufacturing development and 40 percent of the cost growth in procurement. However, it can be difficult to separate cost growth into McNicol's categories. We are indebted to RAND's Greg Hildebrandt for calling our attention to this work by McNicol.

the funding instability measure. And program management resources (leadership, staffs, and funding) may also be associated with funding instability—for instance, under-resourced programs may lack the capabilities needed to explain and defend their activities or to negotiate adjustments that would reduce the adverse effects of funding cuts. As for root causes from outside the Army, one example is geopolitical developments that can dramatically change requirements (such as the end of the Cold War and the 9/11 terrorist attacks). Funding instability may also arise from more-specific strategic and funding guidance from OSD. Army programs are also hostage to developments in national and regional labor and material markets.

As discussed earlier, DoD leaders and analysts typically judge the success of a weapon system program in terms of its cost, schedule, and weapon system performance. Because programs have historically given a high relative priority to weapon system performance over cost and schedule, shortfalls in program management typically take the form of cost and schedule growth from initial plans. Since at least 1960, cost growth and schedule growth have been persistent problems in Army and other programs in DoD despite repeated efforts at acquisition reform.

Rationale for and Selection of Case Studies

Professional intuition and recent empirical analyses provide support for the belief that funding instability should increase cost and schedule growth. But except in some specific circumstances, funding instability may not be the only or even the dominant source of cost and schedule growth. The channels that lead from funding instability to cost and schedule growth are subtle but fairly well defined. As Chapter Two describes, our exploratory analyses of aggregate data for 18 Army programs found only a single association between funding instability and an adverse outcome.[2] Moreover, the sizes of the effects are not well

[2] A larger sample may have been able to detect relationships that were more significant and more subtle.

understood; they depend on what the cause of the funding instability is and on the special circumstances of the particular program. As a result, statistical analysis should not be expected to find a strong correlation between cost or schedule growth and such specific causes as funding instability unless the details of the individual programs being analyzed are captured fairly precisely.

In past efforts, case studies were the preferred analytic approach for teasing out subtle channels of influence in DoD weapon system programs. They thus offered us the best way to pursue our research goals. Case studies have provided much of the knowledge available in the existing literature on the channels of effects relevant to understanding how funding instability affects cost and schedule growth. And because funding instability has not been highlighted in the recent empirical literature, new case studies offer an attractive way to explore this specific issue in detail.

We considered two different approaches for choosing case studies. One approach was to focus on programs with high funding instability and ask what caused the instability and how it affected the management of the programs. The other approach was to look at programs with high and low funding instability and compare the sources and effects of different levels of instability. In coordination with the sponsor, we chose the first of these approaches. Within the parameters of this project, we could conduct only three case studies in sufficient detail to be useful. Because of this small number and the primary policy concern being high funding instability, we chose to focus on recent programs that displayed high levels of funding instability.

We also wanted the cases to be characteristic of current and likely future Army programs. With that in mind, we identified three categories of programs:

1. For expensive systems that the Army will buy moderate numbers of (such as helicopters)
2. For less-expensive systems that the Army will buy very large numbers of (such as munitions)
3. For electronic systems of the kind likely to be increasingly relevant to net-centric operations.

With the sponsor's participation, we then chose one program from each category.

Programs with high funding instability are logical candidates for programs worthy of closer attention. They can be expected to show more pathologies and more opportunities for reactions to changing funding guidance. Table 3.1 extracts the instability ratios for those programs with the highest levels of each measure. For example, the summed absolute value of the differences between planned and actual RDT&E funds for the first five years of the Javelin program, starting at Milestone B, yielded a score of 86 for the planned level of funding at Milestone B during the same period. Measures of instability were even higher for procurement funds. As explained earlier, the total measures of instability are not simple averages of the first two measures of instability.

The purpose of conducting these case studies was to understand the reasons for program funding instability and its effects on program costs, schedule, and system performance. The goal of the entire effort was to learn how program managers react to and deal with problems posed by funding instability. As such, the case studies sought to answer several questions: What caused the funding instability? How did the

Table 3.1
Programs with High Instability, Ordered by Score

Development		Procurement		Total	
Program	Instability Score	Program	Instability Score	Program	Instability Score
Javelin	86	BAT P3I	100	Javelin	81
CH-47F	59	FAAD C2I	100	FAAD C2I	71
FAAD C2I	45	ATIRCM/CMWS	95	BAT P3I	70
BAT P3I	44	JSTARS GSM	93	GMLRS	80
GMLRS	39	SADARM 155	88	JSTARS GSM	47
Bradley Upgrade	37	CSSCS	77	ATIRM/CMWS	43
JSTARS CGS	32	Javelin	75	CSSCS	41
				Longbow Hellfire	41

SOURCE: Calculated from *Selected Acquisition Report* data.

funding instability affect program outcomes—cost, schedule, and performance? How did program managers and Army acquisition leaders cope with and respond to the instability?

The three Army acquisition programs selected as case studies were as follows:

1. Javelin, a missile program that had the highest development and total funding instability
2. FAAD C2I, an electronic system that scored high on all three measures of funding instability
3. CH-47F, a helicopter program with high development funding instability.

These case studies are described in turn in the following subsections. We begin each subsection with a brief description of the system itself, followed by a discussion of our analysis of funding instability for that system. We also discuss the reasons for the instability, the acquisition program's history, and the key events that shaped the program's evolution. Finally, we describe how program managers responded to program and funding instability and set out possible lessons that future program managers may take from the case.

Javelin Case Study

The Javelin missile system illustrates many of the issues that a new weapon system may face as it wends its way through the Army and DoD acquisition process. The first issue is how lengthy the acquisition process can be: The program was born in 1984, but the first test round was not fired until 1993, and the system was not approved for full-rate production until 1997. The second issue is program instability: the Javelin program breached its approved program baseline several times over the course of its development. And the third issue is funding instability. By our measure, the Javelin program had the highest level of research and development (R&D) funding instability and one of the highest scores for procurement funding instability.

System Description

Javelin is a man-portable antitank system (see Figure 3.1) used by both Army and Marine Corps forces. Javelin provides an antitank capability to light, dismounted forces, including infantry, scouts, special forces, and combat engineers.

Javelin has two major components: a reusable command launch unit (CLU) and the missile itself. The CLU incorporates an integrated day/night sight and provides target engagement capability in adverse weather and countermeasure environments. The launch unit also may be used for battlefield surveillance without firing a missile.

The Javelin missile comes packaged with a disposable launch tube that mates with the CLU. Once the operator identifies and selects a target on the CLU screen, targeting information is transmitted to the missile electronically, and the missile's own infrared seeker acquires

Figure 3.1
Javelin Man-Portable Missile System

SOURCE: U.S. Army Infantry homepage.
RAND *MG447-3.1*

the target. The missile is then launched and flies to the target using either a top-attack or direct-attack profile, as selected by the operator. The missile has a range of up to 2,500 meters but can engage a target as close as 65 meters, and it has a tandem warhead designed to defeat both conventional and reactive armor. The key feature of the Javelin is its use of fire-and-forget technology that allows the gunner to take cover immediately after firing the weapon. The missile's seeker continues to image the target during flight even if the target is moving; it also directs the missile to engage the target. Unlike its predecessor, the Dragon missile, the Javelin can be fired safely from an enclosure or a covered fighting position.

Full-rate production began in 1997 with a three-year contract covering FY 1998, 1999, and 2000. Another multiyear contract was signed in August 2000. Over 10,000 Javelin rounds have been produced and delivered to Army and Marine Corps forces. Total acquisition objectives are 20,816 rounds for the Army and 2,553 for the Marine Corps. U.S. Army procurement is scheduled to end with the delivery of rounds authorized and appropriated in the FY 2005 budget. Sales have been made to a number of foreign countries, including Taiwan, Australia, and Jordan, and production for other customers is expected to continue.

Javelin Program Funding Instability

Of the 18 Army programs examined, the Javelin program, at 86, had the highest RDT&E funding instability score (see Table 3.1, above). As described in Chapter Two, RDT&E funding instability is measured by the sum of the deviations from the approved program baseline funding at Milestone B over the first five years of the program. A score of 86 (percent) means that the sum of the deviations was nearly as great as the planned spending itself for that period.

Planned and actual RDT&E spending for Javelin are portrayed in Figure 3.2. Three facts are evident in the figure: The planned development schedule was very short, actual development funding was much greater than planned funding, and actual development took two more years than planned.

Figure 3.2
Planned and Actual Development Funding for Javelin

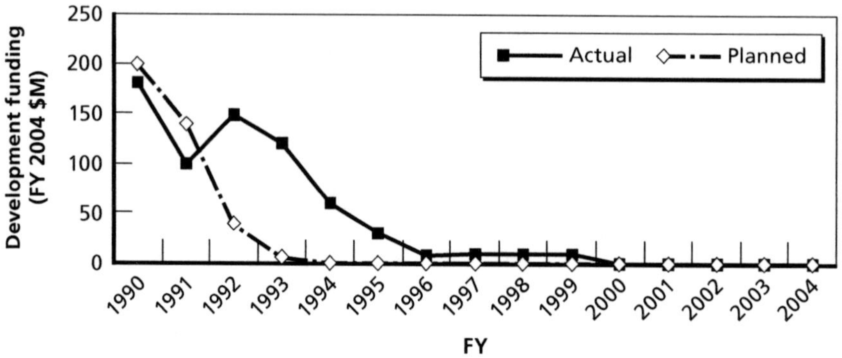

SOURCE: *Selected Acquisition Report* data.
RAND *MG447-3.2*

Javelin's concept development phase, in which three teams of contractors developed proof-of-principal prototype weapons, was unusually long. Approval to proceed to system development (Milestone B) was given in June 1989. The planned system development profile for Javelin had the greatest expenditure, $200 million (FY 2004 dollars), in the first year after Milestone B approval, followed by smaller amounts in the two subsequent years.[3]

Actual development extended through FY 1996, two years longer than planned. And actual development cost for Javelin totaled $964 million (in 2004 dollars), whereas the approved program baseline at Milestone B called for spending $705 million. This is an increase of 37 percent.

At 74, Javelin's funding instability score for procurement was also quite high. Figure 3.3 shows Javelin's planned and actual procurement funding. The path of the planned procurement funding is somewhat unusual—a steeply rising curve for two years of low-rate initial production (LRIP), followed by only four years of full-rate production (FRP) at annual outlays averaging $750 million (2004 dollars). These would

[3] Again, current milestone terminology is being used. See Note 1 in Chapter One for a discussion of milestone designations and definitions used in this monograph.

Figure 3.3
Planned and Actual Procurement Funding for Javelin

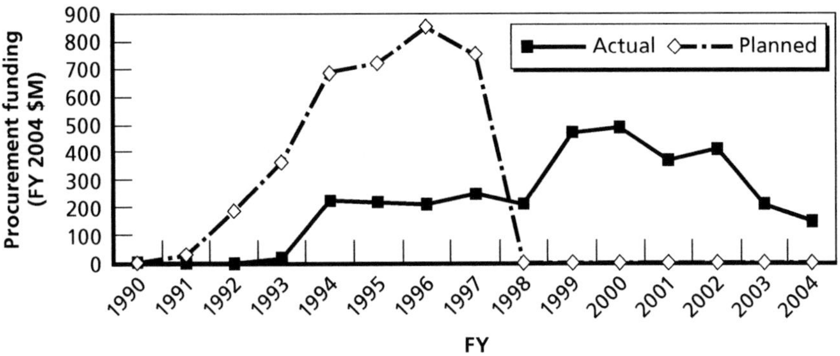

SOURCE: *Selected Acquisition Report* data.
RAND *MG447-3.3*

have been unprecedented levels of annual funding for an Army tactical missile program.

The path shown in the figure for Javelin's actual procurement funding is more normal. As noted previously, low-rate production began in FY 1995, two years later than planned. Three low-rate production buys were executed, followed by eight years of FRP under multiyear contracts. Annual procurement outlays peaked at about $450 million (1997 dollars).

Javelin Program Acquisition History

The Javelin program sprang from Cold War fears of a massive invasion in Europe by the Soviet Union and its Warsaw Pact allies. Warsaw Pact tank forces outnumbered those of NATO by a 3:1 ratio. The fear was that the Soviet tank forces would break through into the open German plain unless there were some other means of keeping them in check. One possibility, the Dragon missile system, was viewed as problematic, and many troops were afraid to use it. NATO commanders viewed a replacement for Dragon, one that could address its shortcomings, as a top priority.

In this environment, on July 11, 1985, the Assistant Deputy Chief of Staff for Operations and Plans, Force Development, Department of the Army, signed the required operational capabilities (ROC) document for the Advanced Antitank Weapon System–Medium (AAWS-M). Responsibility for developing the AAWS-M was assigned to the Advanced Weapon System Project Office, which was also responsible for VIPER (a replacement for the M72 light antitank weapon), the M72E4 product improvement program, and the AT-4 weapon. On September 3, 1985, the Under Secretary of the Army and the Vice Chief of Staff of the Army approved the AAWS-M program for entry into a demonstration and validation phase. Since the AAWS-M program would be a major defense acquisition program if successful, it also required the approval of the Defense Systems Acquisition Review Council (DSARC). A Decision Memorandum authorizing the program office to proceed with the demonstration and validation phase was issued on May 15, 1986.

In August 1986, three contracts were awarded for proof-of-principal technology demonstrations over a 28-month period to Texas Instruments, Hughes Aircraft Company, and Ford Aerospace and Communications Corporation. The results were promising enough that the program office issued a request for proposal in September 1988 for a full-scale development/LRIP (FSD/LRIP) award contract. Approval of the DSARC for entry into Milestone II was granted in June 1989, and the contract was awarded that same month to a joint venture team of Texas Instruments and Martin Marietta Corporation.

The SDD contract schedule was aggressive with respect to both cost and schedule, calling for completion of development and the start of LRIP in 36 months. Table 3.2 shows the schedule milestones.

Then, in January 1991, the contractor team presented a revised estimate of the cost to develop the missile. The cost of the FSD contract—originally targeted at $170 million—was now estimated to exceed $260 million (then-year dollars). As a result, the program executive officer (PEO) ordered that a special "Red Team" be created to review the technical, schedule, and cost status of the AAWS-M program. Cost growth was attributed to various factors, including problems with the focal plane array sensor, problems meeting the target weight threshold, and a compressed test schedule (OSD, 1990, p. 3). After the Red Team reported to the PEO and the Army acquisition

Table 3.2
Schedule Milestone Dates for Javelin

	Development Program		
Schedule Milestone	**Estimated Date**	**Actual Date**	**Schedule Slippage (months)**
Milestone I	May 1986	May 1986	0
Milestone B (II)	Jun 1989	Jun 1989	0
SDD contract award	Jun 1989	Jun 1989	0
Prototype delivery	Apr 1991	Nov 1992	19
Start independent operational test and evaluation (IOT&E)	Jan 1992	Sep 1993	20
End IOT&E	Apr 1992	Dec 1993	20
Milestone C (IIIA)	Jun 1992	Jun 1994	24
LRIP I contract award	Jun 1992	Jun 1994	24
LRIP II contract award	Jun 1993	Mar 1995	21
First LRIP delivery	Sep 1993	Oct 1995	25
Product qualification test (PQT) start	Sep 1993	Apr 1996	31
PQT end	Feb 1994	Jun 1996	28
First unit equipped (FUE)	Feb 1994	Jun 1996	28
Milestone IIIB	Jun 1994	May 1997	35
FRP contract	Jun 1994	May 1997	35
First FRP delivery	Jun 1995	Oct 1998	40

NOTE: Many of the original milestone designations have been retained in this table.

executive, the Army Systems Acquisition Review Council (ASARC) approved a restructured program and a revised 48-month schedule. This plan was presented to the OSD Conventional Systems Committee and the Defense Acquisition Board, and the defense acquisition executive ultimately approved a restructured 54-month SDD program (OSD, 1991, p. 3).

Contribution of Technical Issues to Schedule Slippage

One of the contributors to schedule slippage and cost growth was Texas Instruments' difficulty in manufacturing the focal plane array (FPA) used in the missile seeker. Texas Instruments was a leading maker of semiconductors and had experience in producing FPAs for satellite programs, but it had never sought to mass produce them before. The Texas Instruments approach used a "photo-capacitor" design. Texas Instruments was unable to achieve satisfactory results for yield and production cost. As a result, a government review led to a recommendation to the PEO that the seeker FPA portion of the FPA manufacturing facility at Texas Instruments not be funded.

Hughes Aircraft Company's Santa Barbara Research Center was already under contract to Martin Marietta as a second source using a different FPA design—a "photo-diode." Hughes was directed to accelerate production to provide seekers for the developmentally funded missiles, and its design was ultimately used in the production version as well. Texas Instruments continued to make the CLU's seeker, which was less challenging technically (OSD, 1991, p. 4).

The FPA was only one of the technical challenges for Javelin. One of the program managers cited additional major technical challenges that the Javelin team had to solve:

- A cryogenic refrigerator for the seeker's FPA that would fit on the CLU. The production version was smaller than two D cell batteries, but could cool the array from ambient temperature to the temperature of liquid nitrogen in 2.5 minutes.
- Power supplies. The CLU drained batteries rapidly. ACME Corporation produced a battery that was rechargeable in the field, avoiding a major logistic issue with battery resupply.
- Targeting software that can track a moving target and direct the missile to impact. For security reasons, the missile has to be "dumb" when not in use. The CLU has to download the targeting software to the missile prior to launch, as well as transmit the coordinates of the chosen target. This also makes the software easier to upgrade.
- The rocket motor. Hercules developed the rocket motor, which is currently produced by Aerojet Corporation.

Each of these represented a technical challenge of its own. Together, they presented a task of considerable complexity, and one not likely to be accomplished within the original 36-month development period set by the Army.

Program Baseline Breaches

The Javelin program registered a number of program baseline breaches during development. The first of these stemmed from a relaxation of the original weight goal for the CLU. The revised weight goal was 49.5 pounds, the present weight of the CLU. This program deviation and the revision to the baseline were approved on December 7, 1990, as a result of a Defense Acquisition Board Review.

The second program deviation occurred when the program was restructured in 1991 to reflect the additional costs and schedule revision. This deviation led to a schedule breach, an RDT&E cost breach, and a procurement cost breach. The resulting revised acquisition program baseline was approved on March 30, 1992.

Later in 1992, procurement cost and unit procurement cost breaches to the acquisition program baseline of March 30, 1992, were recorded. These cost increases also caused a Nunn-McCurdy program acquisition unit cost breach.[4] Unit cost increases were attributed to weight-reduction efforts in the CLU housing; increased subcontractor costs; increased system engineering effort; addition of interim contractor support; and new estimates for spares, modifications, depot facilitization, and training equipment. Unit procurement costs were affected by the Marine Corps reducing its purchases by 32 percent, from 12,550 to 8,485, for rounds, and by 38 percent, from 1,486 to 917, for CLUs. The Marine Corps also deferred the start of its production from the LRIP period to the full rate production phase. This had a major effect on near-term production quantities and costs.

---

4 Acquisition law (10 U.S.C. 2433) requires OSD to report to the Congress any time a major defense acquisition program experiences a unit cost increase of 15 percent or more. Such an increase, named for the sponsors of the legislation, is known as a Nunn-McCurdy breach. The increase can be in program acquisition unit cost, which includes R&D cost, or it can be in procurement unit cost. If the increase is 25 percent or more, the Secretary of Defense must not only report the increase, but also certify that (1) the system is essential to the national defense, (2) there is no alternative system that provides equal capability at lesser cost, (3) the new estimate of cost is "reasonable," and (4) the program management structure is adequate to maintain control of costs. (Leach, 2002)

A final Nunn-McCurdy program acquisition unit cost and procurement unit cost breach was recorded in 1993. In this instance, the breaches stemmed from the Army's decision to reduce its purchases of Javelins because of the Cold War's demise and the reduced Army force structure. The Army cut its buys from 58,000 to 26,600 rounds and 5,000 to 2,800 CLUs.[5] At the same time, the Marine Corps made further changes in its planned purchases, reducing its procurement of rounds from 8,485 to 7,011, but increasing its CLU procurement from 917 to 1,055. Table 3.3 summarizes the effect that the end of the Cold War had on Javelin procurement.

After 1993, the Javelin's acquisition path became smoother. LRIP began with the FY 1994 buy. Two additional lots of low-rate production for FY 1995 and 1996 led to the first multiyear production contract, for 1997 through 1999. A second multiyear contract was signed in 2000 for four years. Today, Raytheon Corporation (Texas Instruments' successor) and Lockheed Martin continue to operate the Javelin Joint Venture.

Table 3.3
Effect of Cold War's End on Javelin Procurement

	Number of Units	
Procured Item	**1989**	**1993**
Army rounds	58,000	26,600
Army CLUs	5,000	2,800
U.S. Marine Corps rounds	12,550	7,011
U.S. Marine Corps CLUs	1,486	1,055
Program acquisition unit cost in base-year 1997 dollars:	$56,600	$131,000

SOURCE: *Selected Acquisition Reports* (Office of the Secretary of Defense, 1989, 1993).

[5] The Nunn-McCurdy measures make no allowance for quantity changes when calculating cost growth. Indeed, a significant cut in procurement quantity is often the trigger for a Nunn-McCurdy breach.

Observations from the Javelin Case

Javelin's developmental problems can basically be attributed to two internal factors: the technical complexity of the task the Javelin team took on and the fact that the Army sought to complete development in only 36 months. If the program had been structured for a five-year developmental cycle, the contractor would have had more time to work through the technical challenges, and the funding profile would not have been subjected to such instability.

Javelin's procurement and cost issues basically stemmed from an external factor: the end of the Cold War. The reduction in both the threat posed by Warsaw Pact tank forces and the U.S. Army force structure led the Army and Marine Corps to reduce their planned purchases of Javelin by nearly half. Given the very expensive RDT&E bill for developing Javelin, this led to a major increase in program acquisition unit cost, since the developmental cost would be amortized over many fewer units. Javelin's procurement cost also doubled, meaning that the Army pays as much to acquire 28,800 units as it originally hoped to pay for 58,000.

The Javelin today represents a success story for the Army and provides a much needed anti-armor capability to light or dismounted forces. But its story is a rarity—most programs would have been canceled if they had experienced the technical and cost challenges that the Javelin program had to surmount. And the price the Army ended up paying for Javelin's capability is high compared to what the Army originally planned to pay.

FAAD C2I Case Study

The Forward Area Air Defense System (FAADS) is an integrated network of air defense weapon systems that provides defensive coverage over an Army division. FAADS consists of weapon systems and command and control (C2) elements and is designed to provide battlefield air defense for the Army into the 21st century. OSD first approved the FAADS concept in January 1986. At that point, it consisted of five components, each to be acquired as a separate program, with the

FAAD command, control, and intelligence (C2I) component serving as the link to the other FAADS elements. The five FAADS subprograms were:

1. A line-of-sight forward heavy system for heavy divisions based on an existing air defense vehicle and combining surface-to-air missiles and anti-aircraft guns
2. A line-of-sight rear pedestal mount on the high mobility multipurpose wheeled vehicle to improve Stinger surface-to-air missile effectiveness
3. A non–line-of-sight weapon to counter helicopters
4. Improvements in the anti-air capability of several armored vehicles and helicopters
5. An automated C2I component (the FAAD C2I system).

For this case study, we focused on funding instability in the FAAD C2I system subprogram from 1987 through 1992. The purpose of the FAAD C2I portion of FAADS is to integrate and coordinate the full FAADS program suite. The key FAAD C2I capabilities are

- Collect and digitally process target data
- Disseminate C2I information, including air threat warnings and weapon control orders
- Integrate FAADS into the Army C2 system
- Provide track information to the Combined Arms Initiative (armor, infantry, aviation).

Elements of the FAAD C2I system were initially partitioned into five separate component acquisitions, each with its own tasks, schedule, and budget: (1) FAAD C2, (2) ground-based sensor (GBS), (3) aerial masked target sensor (MTS), (4) non-cooperative target recognition–positive hostile identification (NCTR-PHID), and (5) identification friend or foe (IFF). Funds for all of these elements were included in the original development estimate. Initially, a single FAAD C2I SAR incorporated cost-reporting requirements for all five component acquisitions.[6] Though some components were eventually transferred to other military programs, a single SAR was used for each reporting period in the 1987–1992 time frame.

Our measures of funding instability show high scores for both FAAD C2I development and procurement. In fact, essentially not one of the first five years of planned procurement was accomplished.

Key FAAD C2I Developments

FAAD C2I evolved from the Short Range Air Defense (SHORAD) system's C2 program. In early January 1986, the ASARC directed that SHORAD C2 become part of the FAAD system and be redesignated FAAD C2I. The Joint Requirements and Management Board approved execution of the overall FAAD program at the end of July 1986, approving FSD of the FAAD C2I system software and an acquisition plan for the GBS. A decision on the aerial sensor was deferred until completion of the systems definition phase. The acquisition strategy specified for FAAD C2I relied primarily on non-developmental items and preplanned product improvements to meet system performance goals in a timely manner.

Figure 3.4 shows significant events that affected the FAAD C2I program from 1987 to 1992. As shown, there were two world events: the end of the Cold War in 1989 and Operation Desert Storm in 1991. During the same period, the FAAD C2I program experienced two unexpected problems: the unavailability of government furnished equipment (GFE) in 1987 and the lack of qualified bidders for the GBS solicitation in 1989. In addition, from 1988 to 1992, the program went through three restructurings and a program reduction by Army headquarters. The first restructure, in 1988, aimed to field an early FAAD C2I capability to heavy divisions. In 1990, the focus changed to first fielding a FAAD C2I capability to light and special divisions. Just a year after this change, the Army reduced the program to reflect the smaller post–Cold War force structure. In 1992, the program was

[6] The five separate acquisitions were grouped into three end items for SAR reporting purposes: the FAAD C2, the GBS, and the NCTR, with the NCTR portion encompassing efforts on the MTS, NCTR-PHID, and IFF.

Figure 3.4
Changes Experienced by FAAD C2I, 1986 to 1992

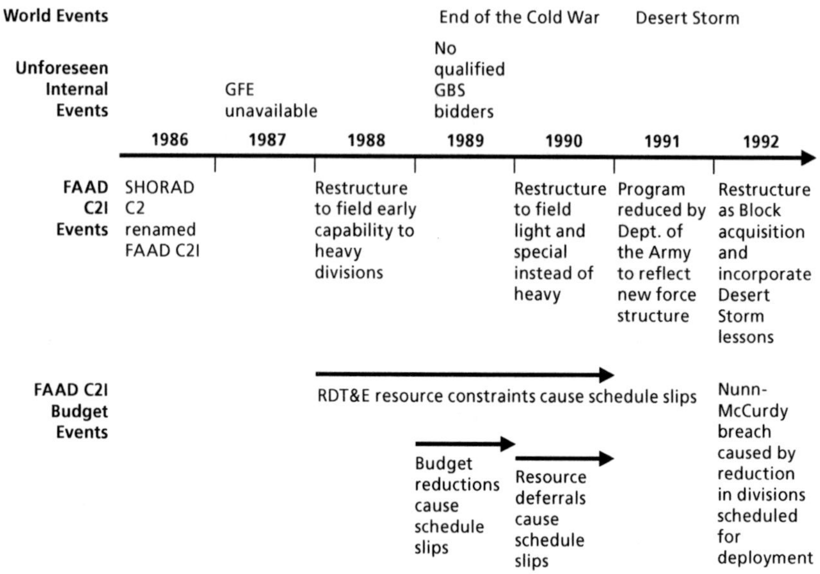

RAND *MG447-3.4*

restructured again, this time to incorporate lessons learned in Operation Desert Storm.

The changes in 1988, 1989, and 1990 caused schedule slips that delayed FUE from 1991 to 1993. As a result of a reduction in the number of divisions scheduled for FAAD C2I deployment, the program experienced a Nunn-McCurdy breach in 1992.

FAAD C2I Funding Instability

The FAAD C2I program experienced no program budget growth from 1987 through 1999. In fact, total expenditures during the period summed to only 74.2 percent of planned expenditures. Total RDT&E spending during the period was close to the plan; but because of the various program restructurings, procurement summed to only about 50 percent of what had originally been planned.

The program experienced high funding instability. Using our instability measure, the FAAD C2I program had an RDT&E funding instability score of 45 and a procurement instability score of 100. Figure 3.5 shows how actual year-by-year expenditures differed from planned expenditures for RDT&E; Figure 3.6 shows the same thing for procurement. The following sections detail the events that appear to have caused the high funding instability in the FAAD C2I program.

Figure 3.5
Planned and Actual RDT&E Funding for FAAD C2I, 1987 to 1999

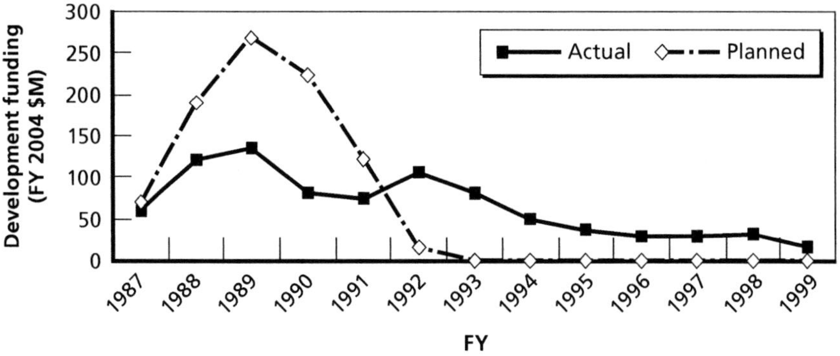

SOURCE: *Selected Acquisition Report* data.
RAND *MG447-3.5*

Figure 3.6
Planned and Actual Procurement Funding for FAAD C2I, 1987 to 1999

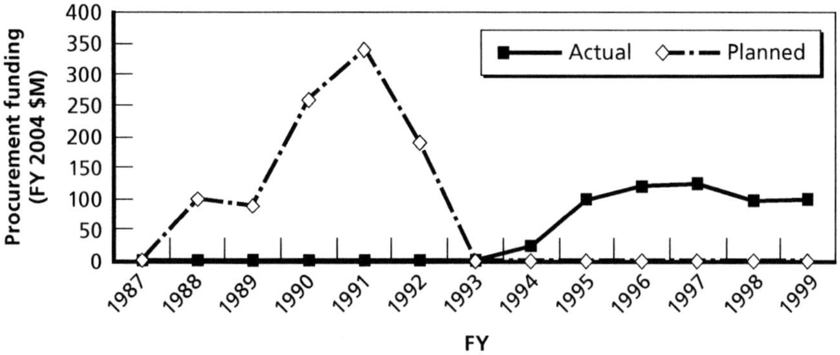

SOURCE: *Selected Acquisition Report* data.
RAND *MG447-3.6*

Key Events Impacting Funding

1987. Two factors—delayed availability of GFE and budget reductions—explain the difference between planned and actual funding for the FAAD C2I in 1987.

The GFE deliveries delayed were the Joint Tactical Information Distribution System (JTIDS), Enhanced Position Location and Reporting System (EPLRS), and Global Broadcast System radios needed to develop and test the FAAD C2I system software. The FAAD C2I team needed these types of radios for FAAD systems development. At the time, these radios were still under development, and the FAAD C2I project intended to use development models provided through the government project managers. Although the JTIDS radio was a joint item, the FAAD C2I team needed a special Army version that would be used only in air defense. (The Air Force JTIDS used a data bus that was not compatible with ground vehicle electronics.) This special model was only to be produced in small quantities (whether this contributed to its delayed availability is unclear). The JTIDS radios cost approximately $1 million each (in FY 1987 dollars); several sources cited this expense, but we found no evidence that the unit cost caused schedule changes that resulted in delivery delays. Though we could not determine the root cause of the JTIDS and EPLRS schedule changes, both projects experienced schedule modifications that delayed delivery of the radios to the FAAD C2I project.

Army headquarters made the budget reductions, but we could find no documentation explaining the reasons for them. Thus, there is no evidence that the budget actions were caused by the delayed availability of the GFE. Recollections of current FAAD C2I personnel suggest that the budget actions were independent of the delayed GFE availability.

Tracing the two factors—delayed availability of GFE and budget reductions—through the 1987 FAAD C2I SAR provides insight into how the 1987 budget instability occurred. The December 1987 SAR states that schedule slips (ranging from six to 28 months depending on the specific activity) were caused by delayed availability of GFE and budget reductions. The same SAR attributes a $22.54 million decrease to schedule changes, which accounts for more than 99 percent of the difference between the planned and actual 1987 funding. Hence, the

1987 FAAD C2I budget instability appears to have stemmed from delayed availability of GFE and budget reductions. These events jointly led to schedule slippage, and the schedule slippage resulted in a total financial impact of $22.54 million.

1988. The Chief of Staff of the Army requested a FAAD C2I program restructure, and the Deputy Secretary of Defense approved it in November 1988. The restructure aimed at fielding an initial capability for limited air defense engagement and essential force control interfaces in a division. The previously planned automated force control software became a follow-on preplanned product improvement. These changes primarily addressed the portion of the program most affected by the delayed availability of GFE and aimed to mitigate the effects of the GFE problem.

The decision approved the fielding of an initial capability to perform limited air defense engagement and essential force control interfaces within a division. Actual implementation of the restructure involved transferring parts to make a fieldable system for heavy divisions. The transfer of activities caused the delay of some activities and the acceleration of others that had originally been scheduled for later years. Still other planned activities were canceled altogether because the restructure did not include them.

The delayed availability of GFE in 1987 and the resultant schedule slippage were the primary causes for the restructure. This restructure accounts for the entire difference between actual and planned expenditures in 1988. Thus, the delayed availability of GFE remains the root cause of the 1988 funding instability.

1989. Together, the 1988 program restructure, a failed solicitation for the GBS, and an administrative transfer of IFF funds to the Air Force explain the difference between planned and actual amounts in 1989.

The first of these, the 1988 program restructure, was not implemented until 1989. This lag explains some of the difference between the 1989 planned and actual amounts.

The second cause of some of the difference was the failure of the FAAD C2I program solicitation for the GBS to yield any qualified bidders. In consequence, none of the planned procurement expenditures was incurred in 1989, and funds earmarked for GBS contract award were withdrawn. Some GBS RDT&E funds were reprogrammed by Army headquarters to other, higher-priority uses while a new GBS acquisition strategy was devised.

A survey conducted following the failed GBS solicitation revealed that the Army's requirements had been too stringent. As a result, the requirements were revised and a different acquisition strategy was adopted in an effort to attract qualified bidders. Thus, the cause of these changes can also be considered internal—i.e., stemming from an overly ambitious solicitation. Had this survey been conducted prior to the initial solicitation, the solicitation's requirements and acquisition strategy might have been revised to good effect.

The third item contributing to the difference was the transfer of the IFF development efforts and their associated funds to the Air Force. Our funding instability measures cannot account for changes in program scope. The planned IFF funds were included in the planned Army amounts, but IFF expenditures were no longer included in the actual Army spending.[7] So this shift of funding adds to the FAAD C2I funding instability scores. This shift is an administrative change external to the FAAD C2I program.

1990. The failed GBS solicitation and a program redirection are the primary explanations for the difference between planned and actual funding in 1990. (The removal of IFF funds also continued to account for a portion, a small one, of the funding difference.)

The 1990 planned procurement spending assumed that the 1989 GBS solicitation would result in a contract and GBS buys. Because there was not a qualified bidder, the 1990 procurement funds were not spent, except for a minor amount. The failed solicitation led to funding reductions.

The planned RDT&E for 1990 was sharply reduced. In April 1990, the Army acquisition executive approved early development of a tailored FAAD C2 for fielding to light and special divisions, a redirection that was approved by OSD in July 1990. This redirection required that a light forces FAAD C2I capability be developed using interim

[7] They are, of course, reported in Air Force accounts.

sensors, communication systems, and electronics tailored for light and special division needs. The interim components were largely available from industry. This restructure reflected the changing threat and budget reductions, as well as the availability of components for light systems.

The combination of the failed GBS solicitation and the program refocus led to program stretch-outs and changes in the fielding schedules. The stretch-outs incurred fixed costs in later years; the fielding schedule changes affected the amounts required because the costs to field light and special divisions differed from the costs to field heavy divisions.[8] The difference between actual and planned amounts in 1990 is primarily the net financial impact of the failed GBS solicitation and redirection of the program.

1991. The Army's reduction in the FAAD C2I program and the failed 1989 GBS solicitation continued to contribute to funding instability in 1991. Beyond those earlier events, an Army-directed shift of computer expenditures from procurement funds to operations funds, the deletion of GBS RDT&E funding, and some changes in program scope added to the difference between planned and actual funding.

None of the planned $336.4 million procurement was executed. Taken together, four changes explain the difference in the planned and actual procurement amounts in 1991.

First, the Army-directed program reduction provided for the purchase of FAAD C2I capabilities for forces deployed overseas and rapid deployment forces in the continental United States (CONUS), as well as for a training base. The Department of the Army had previously planned to buy an additional package for heavy divisions and other forces.

Second, the NCTR portion of the FAAD C2I program was reduced to reflect the newly defined force structure.

Third, the planned procurement amount had included funds for buying the GBS. However, no GBS procurement occurred at this time

---

[8] The cost of fielding a heavy division was estimated to be about three times the cost of fielding a light or special division, per consensus of expert opinion of personnel involved in the FAAD C2I program during the 1987–1992 period.

because the second GBS solicitation did not result in a contract award until 1992.

Finally, in 1991, Army headquarters directed that expenditures for computer buys that had been covered by procurement funds be shifted to operations and maintenance funds, which are not reported in the SARs. This amounts to a change in accounting similar to that associated with the earlier shift of IFF work (and related funds) to the Air Force. Hence, the amounts expended in 1991 for computers were not reflected in the actual amounts for 1991.[9]

The planned RDT&E amount for the FAAD C2I program in 1991 was $114.0 million, and the actual amount was $78.4 million. RDT&E for the FAAD C2 portion of the program accounted for $67.4 million of the actual amount, with the remainder spent on RDT&E for NCTR. The 1991 RDT&E funds for GBS were reduced to zero, but GBS RDT&E was funded with a carryover of about $9.1 million (in then-year dollars) of FY 1990 RDT&E funds.[10] The program also received $14.8 million in FY 2004 dollars in Foreign Comparative Testing Funding. (However these GBS RDT&E expenditures are not included in the SARs and therefore are not included in the actual RDT&E expenditures for 1991.)

Before September 1991, the SARs included development funds for the MTS. As of 1991, however, funding and work on the MTS ceased, and actual amounts for 1991 and beyond no longer included expenditures for the MTS. This is another change in scope. The development planning estimates still included MTS funds.

Finally, there were initially six NCTR devices. However, the actual amounts reported in SARs dated September 1991 and beyond no longer included expenditures for one of the six because it was no longer a system designated for FAADS. In addition, two other NCTR devices were not funded in 1991. Hence, the planned amounts included

---

9 This action was reversed in 1995, when computer re-buys were added back into the Other Procurement, Army (OPA) requirements.

10 How much of the 9.1 million then-year dollars are FY 1988, 1989, and 1990 dollars is unclear, so an exact equivalent of FY 2004 dollars cannot be calculated with the limited information currently available.

funds for the six NCTR devices, but the actual amounts do not include expenditures for three of the six.

The convoluted 1991 GBS RDT&E funding scheme, the ceasing of MTS work, and the change in scope of the NCTR effort together account for the difference between the planned and actual amounts in 1991 RDT&E funds.

1992. Differences between planned and actual amounts in 1992 had two sources. A third FAAD C2I program restructure decreased the number of divisions scheduled to receive FAAD C2I capabilities. Additional changes led to expenditures to incorporate lessons from Operation Desert Storm. The sources of these causes were external to the FAAD C2I program.

The FAAD C2I program was restructured to a block acquisition format. The restructured program included reductions to reflect a decrease, from 35 to 19, in the number of divisions scheduled to get the system, and added allowances to incorporate lessons learned in Operation Desert Storm. With the restructure, cost reporting changed to two instead of three end items, a change that when coupled with the reduced number of divisions scheduled for deployment, produced a program acquisition unit cost breach (Nunn-McCurdy). The December 1992 SAR established new unit cost baselines for the two new end items, identified as Block I (light and special divisions) and Block II (other divisions and forces).

Observations from the FAAD C2I Case

Our review found little evidence that budget changes made by Army leadership were responsible for the high FAAD C2I budget instability score. Rather, two major external events, the end of the Cold War and Operation Desert Storm, were the root causes of much of the funding instability. Funding instability was also caused by the unforeseen delays in GFE availability and the lack of qualified bidders in the first GBS solicitation. The resulting three program restructures, all of which aimed to mitigate the negative effects of these events, were the primary causes for the high instability scores. More minor contributions to the instability score came from these sources:

- An independent budget reduction by the Department of the Army in 1987
- The shift in IFF funding from the FAAD C2I program to the Air Force in 1989
- The zeroing of GBS RDT&E funding in 1991
- Changes in the scope and cost reporting of effort in the MTS and NCTR portions of the program in 1991
- The funding change for ten-year computer buys from procurement to operations funding in 1991.

Chinook Improved Cargo Helicopter (CH-47F) Case Study

The CH-47F program, initiated in 1996, is intended to modernize a large portion of the Army's fleet of cargo helicopters, which currently consists of over 400 CH-47D Chinook aircraft. In addition to producing a new F-model of the Chinook, the program will produce 37 MH-47G aircraft, helicopters specially modified for Special Operations Forces (SOF).

CH-47 System Description

The CH-47D is the Army's cargo helicopter for heavy lift missions. The Chinook has twin turbine engines and two rotor assemblies, and its principal mission is to move troops, artillery, supplies, and equipment around the battlefield. It may also be used for such missions as medical evacuation, aircraft recovery, parachute drops, and search and rescue.

Chinook aircraft are built, modernized, and supported by the Boeing Corporation at its plant in Philadelphia. Boeing delivered the first Chinook to the Army in 1962; A-, B-, and C-model Chinooks served throughout the Vietnam War. After the war, the Army directed Boeing to undertake a major modernization program for the Chinook. In 1982, Boeing delivered its first D-model aircraft; it concluded that program in 1994. Only two U.S. Army CH-47Ds were built new to replace aircraft lost in Operation Desert Storm. All other D models are remanufactured A-, B-, and C-model Chinooks.

Compared with its predecessors, the CH-47D features composite rotor blades, an improved electrical system, modularized hydraulics, triple cargo hooks, avionics and communication improvements, and more-powerful engines that can handle a 25,000-pound useful load (nearly twice the Chinook's original lift capacity). The Chinook has been the U.S. Army's prime mover for 20 years. It was a central element in U.S. Army operations in Operation Desert Storm, where more than 160 Chinooks carried U.S. and allied troops in history's largest aerial assault to outflank Iraqi forces and cut off their retreat from Kuwait.

The 160th Special Operations Aviation Regiment (Airborne) operates specially modified Chinook aircraft designated as MH-47Ds and MH-47Es. Twelve MH-47Ds and 26 MH-47Es were produced by converting and equipping CH-47D airframes. The MH-47Es are among the most advanced rotorcraft in operation today. They incorporate fully integrated digital cockpits; forward-looking infrared, terrain-following/terrain-avoidance radar; long-range fuel tanks; and aerial refueling capability. SOF Chinooks perform low-level, high-speed flight for infiltration and retrieval of SOF teams at night and in adverse weather.

CH-47F Upgrade Program

The CH-47F program is designed to extend the service life of the current fleet of CH-47D aircraft by an additional 20 years. The CH-47F incorporates improvements to

- Enhance airframe reliability and maintainability to reduce operating and support costs
- Upgrade the avionics suite to comply with the standards contained in DoD Joint Technical Architecture–Army
- Increase the aircraft's lift capacity and range
- Add support for the Air Warrior aviator ensemble
- Comply with Global Air Traffic Management and Digital Source Collector "flight data recorder" requirements.

New T55-GA-714A engines will be installed to provide the improvement in range and lift capacity. Most of the subsystems currently on the CH-47D will remain, with the program replacing 97 dynamic components with new hardware and repairing or replacing remaining parts, as required, after disassembly and inspection.

The CH-47F program, when completed, will have delivered 300 modernized CH-47F cargo helicopters.[11] The modernization process involves stripping a CH-47D aircraft down to its bare metal airframe, repairing any worn or damaged parts, stiffening the airframe to reduce vibration, and equipping the aircraft with new wiring, plumbing, and systems. The result will be a zero-hour aircraft capable of performing for at least 20 more years.

In addition to producing the 302 CH-47Fs, the program will manufacture 37 MH-47Gs. The G-model aircraft, which is specifically for SOF, will essentially combine the specially equipped cockpit and avionics found in the MH-47E model with the strengthened airframe and improved power train of the CH-47F.

Table 3.4 presents the CH-47F program's key schedule milestones, highlighting dates that slipped or were changed. The reasons for the changes are discussed next.

System Development and Demonstration Phase[12]

The CH-47F modernization program was approved for SDD in November 1997 by ASARC. In March 1998, a cost-plus-incentive-fee contract was awarded to the Boeing Helicopters Aircraft Systems Division for the development of the CH-47F and for production of two aircraft for testing. The negotiated cost was $67.5 million, and the target price—including an estimate of the incentive fee—was $76.1 million.

By DoD standards, cost growth for the CH-47F program during the development phase was not excessive. As of the December 31, 2001, SAR, the Boeing contract cost had risen by $11.8 million, or 16

---

[11] Reportedly, the Army is thinking about increasing the number of CH-47Fs it will produce, and may make some of them through new production rather than conversion of a CH-47D aircraft. (Interview at Cargo Helicopter Project Office)

[12] This discussion uses then-year dollars for all contract costs.

Table 3.4
Schedule Milestones for CH-47F

Schedule Milestone	Development Estimate	Current Approved Program	Actual or Latest Estimate
Approval of operational requirements document	Nov 1997	Nov 1997	Nov 1997
Milestone B (ASARC)	Nov 1997	Nov 1997	**Dec 1997**
SDD contract award	Mar 1998	Mar 1998	**May 1998**
Contractor design review	Sep 1999	Sep 1999	Sep 1999
LRIP contract award	Dec 2001	**Dec 2002**	**Dec 2002**
Complete IOT&E	Mar 2002	**May 2004**	**May 2004**
First LRIP aircraft delivered	May 2003	**Oct 2004**	**Oct 2004**
Milestone C (ASARC)	Jan 2004	**Nov 2004**	**Nov 2004**
FRP Contract Award	Feb 2004	**Dec 2004**	**Dec 2004**
FUE	Sep 2004	**May 2007**	**Nov 2007**

SOURCE: *Selected Acquisition Report* (Office of the Secretary of Defense, 2003).

NOTE: Dates shown in bold type are those that slipped or were changed.

percent, and overall RDT&E cost was estimated to have increased by $14.4 million. The December 2001 SAR narrative attributes the bulk of this increase to a change in the contractor labor rate. By the end of 2003, RDT&E costs had risen by a cumulative total of $21.4 million, $20 million of which was attributed to estimating changes (higher labor and other contractor costs).

Our statistical analysis of instability recorded a score of 59 during the SDD phase for the CH-47F program. This was one of the higher scores for the programs in our sample. Figure 3.7 compares planned and actual RDT&E funding for the CH-47F program.

Schedule Adjustments

As can be seen in Figure 3.7, the measured instability basically resulted from a slip of one to two years in the planned program. This is confirmed by the program's revised milestone dates, shown above in Table 3.4. The major perturbation to the development program was actually the Army's decision to delay the start of LRIP by one year based on the program manager's recommendation. As a consequence, additional funding was needed to continue to support the development effort for an additional year. Another complicating factor was that the unit des-

Figure 3.7
Planned and Actual Research Funding for CH-47F

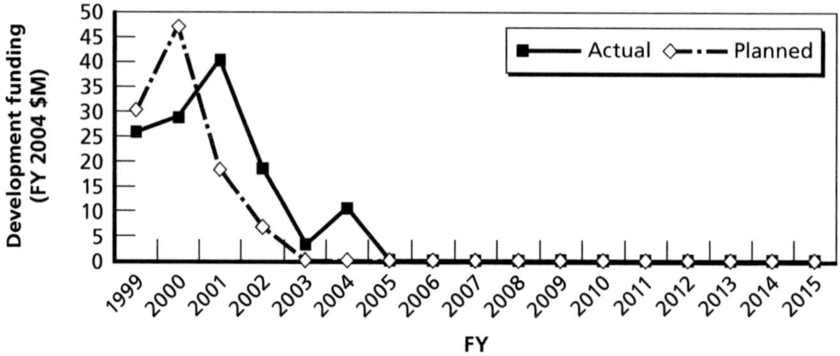

SOURCE: *Selected Acquisition Report* data.
RAND *MG447-3.7*

ignated to perform the initial operational test was deployed to Operation Enduring Freedom, necessitating that the test be delayed.

The approved developmental program baseline called for LRIP to begin in FY 2002 with an initial lot of seven aircraft. Funds had been provided in the FY 2002 President's Budget (submitted to Congress in March 2001) to support that decision. And the LRIP contract to procure the first lot of aircraft was to be signed by December 2001.

Three intervening factors, however, caused the program manager to rethink the plan, and all of them were related to cost:

1. Higher contractor labor and overhead rates
2. Higher than expected over-and-above costs (i.e., costs that are incurred to make necessary repairs to the airframe and that cannot be determined until aircraft are disassembled)
3. Costs associated with full component recapitalization (i.e., replacement of parts with new ones even when they were unchanged for the new model) and other changes to the remanufacturing plan.

In late 2001, Boeing revised its estimates of labor and overhead rates for the program, significantly increasing them. The result was an average increase of about $1.9 million for each aircraft, or a total of $580 million in then-year dollars over the life of the program.[13]

Boeing also reported that the over-and-above costs for the two aircraft being produced under the development contract were running significantly higher than had been projected. The revised estimate for these costs added $1.8 million per aircraft.

In 2001, the Vice Chief of Staff of the Army directed the program office to incorporate a number of improvements into the CH-47F that were not part of the original statement of work: full component recapitalization, global air traffic management systems needed for civil airspace operations, the Air Warrior ensemble, and a digital flight data recorder (Digital Source Collector). Full component recapitalization was estimated to add $3.4 million per aircraft.

Together, the increases added over $7 million to the cost of producing each aircraft—they effectively doubled procurement costs. In consequence, the funds that had been appropriated for the start of LRIP could purchase only three aircraft, not the seven planned.

Starting a production line to produce only three aircraft over an entire year did not make sense to the program manager, so he sought the Army acquisition executive's approval for a one-year slip in the program. That approval was given at a special ASARC meeting on March 7, 2002. The revised program was formulated in time to be presented to Congress in the SAR for December 2001. The Under Secretary of Defense for Acquisition and Technology notified Congress of the Nunn-McCurdy breaches in program and procurement unit cost on May 2, 2002.

Factors Causing Production Instability

Congress approved the Army's request to buy seven aircraft in FY 2003, and the LRIP contract for those aircraft was awarded in December 2002. Another 16 aircraft were funded in FY 2004, and a third lot, also of 16, was included in the President's FY 2005 budget. The LRIP rate of 16 aircraft (four deliveries every quarter) was not all that differ-

[13] These amounts and those in the following paragraphs are all in program base-year 1997 dollars.

ent from the anticipated full rate of 23 to 26 aircraft a year (about six deliveries every quarter).

We measured production instability for the CH-47F program for the first five years of planned procurement, FY 2001 through FY 2005.[14] Figure 3.8 shows the planned and actual funding for the period. The procurement instability score for the CH-47F program was 31. What factors were responsible for that instability?

Two events had a major impact on the CH-47F program. The first was the major increase in aircraft manufacturing costs discussed above. That increase is evident in Figure 3.8 in the increased cost of buying similar quantities of aircraft in FY 2004 through FY 2006. Also evident in the figure is the one-year slip, from 2002 to 2003, in the start of procurement.[15]

The second major program impact stemmed from the decision to give priority to the production of MH-47G aircraft for the SOF. This choice reflects the initial experience of the war on terrorism in Afghan-

Figure 3.8
Planned and Actual Procurement Funding for CH-47F

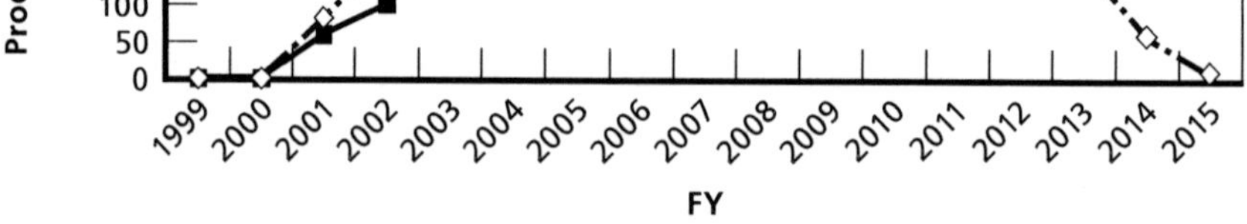

SOURCE: *Selected Acquisition Report* data.
RAND *MG447-3.8*

[14] Although production was to begin in FY 2002, long-lead procurement money was programmed in FY 2001.

[15] As already noted, procurement funds in 2001 and 2003 paid for long-lead items, as well as nonrecurring facilitation costs at Boeing and its subcontractors.

istan. Six of the seven aircraft authorized in FY 2003 were MH-47Gs, as were all 16 aircraft authorized in FY 2004 and 14 of the 16 in FY 2005. As a result of that decision, the first (CH-47F) unit would not be fully equipped with aircraft until November 2007. This is a delay of 38 months from the date set out in the developmental baseline—an extremely long schedule slip, especially for a program of this magnitude. However, it is a somewhat artificial "slippage," since two units of SOF will be fully equipped before the official "first unit" receives its complement of cargo aircraft.

Although the decision to give priority to MH-47G production certainly affected the schedule for CH-47 production, it did not have a major fiscal impact on the CH-47F program. The MH-47G is a more expensive aircraft than the CH-47F version because of all the advanced avionics and communications equipment it carries, as well as the modifications it needs for longer-range operations (air refueling capability and extended-range fuel tanks). It is not possible to say how much more expensive it is, however, because of the way these modifications are funded. The Army Materiel Command and the Cargo Helicopter Program management office are only responsible for funding the elements the two helicopter models share, which represent about 86 percent of the cost of a CH-47. The remainder of the cost of an MH-47G is funded from Special Operations Command (SOCOM) procurement appropriations and managed by the SOCOM's technical applications program office.[16] For this reason, the decision to produce MH-47Gs first did not impose any fiscal burden on the Army.

Observations from the CH-47F Case

In the case of the CH-47F program, funding instability was caused by both external and internal factors, chiefly the increase in the contractor's labor costs (an external factor), and the increase in over-and-above costs and full component recapitalization costs (two internal factors). These factors most heavily impacted the production funding instability.

---

[16] Examining the details of SOCOM-managed activities was beyond the scope of our project.

Part of the instability is attributable to the program manager's decision to slip the start of LRIP by a year. This especially affected the developmental program, which up to that point had proceeded according to schedule.

One could come to the conclusion that the CH-47F program is atypical. Its development phase only cost $160 million, which is a small price for a major defense acquisition program. And to a greater extent than many programs, the instability during the development phase could be characterized as self-inflicted.

However, the CH-47F case is instructive of how—faced with a crisis—a program manager can respond with appropriate actions to steer his program toward achieving its ultimate goals. In that sense, it is a good case study of how a program manager responds when outside factors create problems and obstacles that threaten his or her program.

What the Three Case Studies Reveal

The Javelin program began with an aggressive schedule and cost plan but was unable to overcome development problems within the planned schedule. When confronted with its first major budget instability, the product of primarily technical problems and an overly optimistic schedule, program management responded by seeking a longer, revised schedule more consistent with the program's technical complexity. When the end of the Cold War led to government-imposed reductions in procurement quantities, program management sought and executed multiyear contracts to minimize any further unit cost increases.

The FAAD C2I program began with realistic schedules and budgets, and then was forced to traverse a volatile path by external and unexpected events. Upon learning that GFE would be delayed and the Army was cutting the program's budget, program management sought to restructure the program to minimize the effects that the delay and budget reduction would have on fielding of the capability. Shortly after this first program restructure, management was surprised with a failed GBS solicitation; it responded with a second program restructure, this time to minimize the effects of the unexpected failure. The next turn of events was the end of the Cold War, which led to reductions in the U.S. force structure that, in turn, led to government cuts in the FAAD C2I program. Shortly after program management learned of this reduction, Operation Desert Storm concluded; and with its end came a golden opportunity to incorporate lessons learned into the FAAD C2I system. Program management instigated a third program restructure to absorb the program reduction and allow incorporation of lessons learned in Operation Desert Storm.

The CH-47F program encountered unexpected cost increases shortly after program initiation. Program management responded proactively, requesting a delay in LRIP to better realign the program with the planned funding profile. When program management was directed to change its production plan and give priority to SOF aircraft, it realized that the financial impact would be minimal and simply negotiated a new date for the first unit to be equipped. In other words, when confronted with program perturbations, CH-47F management responded with different actions to ensure that the program's goals would be met.

Together, these three case studies reveal that funding instability is most often a consequence of changes occurring outside the Army. The end of the Cold War and the resulting force reductions are a clear example of such changes. But other factors, such as contractor cost increases and the need to adopt lessons learned from recent operations, are externally driven imperatives that can lead to funding instability. Within the Army, the approval of ambitious plans can create funding instability by leading to delays and slippage in development and procurement program schedules. Nonetheless, our three cases indicate that most funding instability in Army programs is the consequence of external changes rather than reallocation of funding among Army programs. And regardless of the what causes funding instability, its effects show up as schedule slippages, cost increases, and, to a lesser degree, technical compromises.

CHAPTER FOUR

More-Recent Funding Instability

Chapters Two and Three describe our analysis of funding instability in a variety of Army programs active in the 1990s. A few of those programs were long-lived ones that had begun in the 1970s. The oldest, the UH-60A Black Hawk helicopter, had its Milestone B review in August 1972.[1] Most programs included in the database had their Milestone B decisions in the 1980s or 1990s, however, with the latest being the GMLRS, whose Milestone B decision was in November 1998. Two of the three case study systems—Javelin and FAAD C2I—accomplished Milestone B in the 1980s. Thus, the experience that we analyzed may not represent more-recent acquisition priorities, processes, and policies. To provide a more recent perspective on funding instability, this chapter describes the analysis of funding instability experience in Army and Air Force programs from 2000 through 2004 that we carried out to examine the concerns of Army policymakers.

Army leaders have expressed concerns that funding cuts and reallocations have harmed major acquisition programs. This perception was likely shaped by more-recent, rather than earlier, experiences. The funding of transformation activities required that Army funds be shifted, particularly among R&D programs. Moreover, recent experience includes the arrival of a new administration, the 9/11 attacks, and the urgent requirements of the Global War on Terrorism. As a consequence, the lessons of earlier periods may have been "overtaken

[1] Approval to begin system development was called Milestone II until 2000, at which point it became Milestone B. See Note 1 in Chapter One for a discussion of milestone designations and definitions and the use of those designations in this document.

by events," and significant instability in RDT&E and procurement programs might be expected in 2000 through 2004.

In this section, we use more-recent data to address two issues. First, we examine whether Army programs experienced higher levels of funding instability in 2000 through 2004 than they did in the 1980s and 1990s. For this examination, we estimated funding instability measures for 16 Army programs over five years, FY 2000 through FY 2004. Ten of the 16 programs were in the original set of 18 programs used in our previous calculations. The six programs not common to the original set were those for the Abrams tank upgrade, Comanche helicopter, Force XXI Battle Command Brigade and Below (FBCB2), Patriot Advanced Capability 3 (PAC-3) fire unit, PAC-3 missile, and Stryker (all of which are described in Appendix A). The baseline for measuring instability was the program funding plan baseline contained in each program's December 1999 SAR. The estimation procedures we used were identical to those described in Chapter Two.

Second, we examine whether recent funding instability was any greater for Army programs than for Air Force programs. A report on one study of Army acquisition management stated that "Army programs appear to suffer even more budget instability than do acquisition programs in the other Services" (Center for Naval Analysis, 2001, p. 5). The broad impact of an administration change affected all services, however, and each service felt pressure to speed transformation. In addition, all of the services were required to adapt to the needs of the Global War on Terrorism, so a comparison of the two services' experiences is one way to help identify whether Army funding instability has been greater.

It needs to be noted, here, that the data on FY 2000–2004 programs differ in some ways from the data used in our exploratory analysis. The most important difference stems from the fact that the earlier analysis used information for the first five years after Milestone B, which means that the programs were at the same stage of their life cycle but that their defense planning and fiscal environments varied. One of the programs in that analysis had passed its Milestone B in the 1970s, but the other 17 had been given Milestone B approval after FY 1985. Defense Department and Army spending (in constant dollars) peaked in FY 1985 and declined continuously through FY 1998. In general, budgets were being cut and program plans were implemented in a period of increasing fiscal stringency.[2]

For our analysis of the more recent programs, the data we used were all from FY 2000 through FY 2004, so all programs functioned in a common planning and budget environment. The fiscal environment was significantly different from the environments in the earlier periods. DoD and Army spending plans reflected increasing availability of funds for acquisition programs. Unlike the programs used in the earlier analysis, however, the individual programs examined here were at different stages of their life cycle.

Examining comparative funding instability, though not a part of the project's charter, became feasible for two reasons. One was the project's new measure for funding instability, and the second was the availability of comprehensive data from more-recent SARs. To see whether the Army experienced greater funding instability than another service did, we compiled comparable funding instability measures for 19 Air Force programs active in FY 2000 through FY 2004 and compared them with those for Army programs during the same period.

Methodology for Comparisons

The funding instability measures for the more recent programs were developed from data reported in the December 1999 and December 2003 SARs. The December 1999 SARs contained the projected RDT&E and procurement spending for FY 2000 through FY 2004 that was consistent with the Clinton administration's FY 2001 budget request and the FY 2000–2005 Future Years Defense Program. Those estimates represented the program baseline as of December 1999. To estimate funding instability, we compared these projections with the actual RDT&E and procurement funding for the period, as reported in the December 2003 SARs. As already noted, individual programs in

2 Three of the Army programs—FMTV, CH-47F, and GMLRS—achieved Milestone B in FY 1998, so their plans were executed in a period of growing Army budgets.

the December 1999 SAR were at varying points of their development and procurement efforts.

Both sets of numbers were adjusted to constant 2004 dollars. Then the year-by-year projected spending was subtracted from the actual funding, and the resulting differences were converted to absolute values.[3] Funding instability is measured by dividing the five-year sum of the absolute differences by the planned program's five-year sum of spending and expressing the result as a percentage. Separate calculations were made for RDT&E and for procurement funding. (See Chapter Two, Figure 2.4, for an illustration of the calculation.)

For some of the programs, data were missing: Either no planned or no actual data were available for some or all of the 2000–2004 period. (This might occur, for example, if there were no December 2003 SAR because the program had been canceled.) When a valid comparison could be made for an earlier—but still recent—five-year period, such as FY 1999 through FY 2003, the funding instability data for that period were substituted in the analysis. In some instances, however, no score could be computed because no spending had been planned.

The resulting measure is not directly comparable with the funding instability scores defined earlier, in Chapter Two. Those scores were computed for the five years immediately following the key program milestone, either Milestone B, the start of system development, or Milestone C, the start of procurement. Thus, those R&D and procurement estimates were typically based on different sets of fiscal years.

The more recent results were computed for a common set of fiscal years without regard to the timing of program milestones. If programs had greater funding instability in their first five years than later, the latter estimates of funding instability could be expected to be lower. On the other hand, if recent political-military events had led to multiple, continuing changes, the latter estimates could be higher.

[3] The implication of adding the absolute values of funding changes is that both cuts and increases affect program management. Both require managerial changes, such as new budgets and plans, as well as new contracts and negotiations. But the effects of cuts and increases may not be symmetric, particularly when the amount of funding instability is large relative to the size of the planned program.

Has Funding Instability in the Army Increased?

Research and Development Programs

Table 4.1 shows financial instability scores for the more recent Army R&D programs. As can be seen, scores for the FY 2000–2004 period ranged from 8 (Longbow Apache program) to 212 (GMLRS). The dollar amount reported in the table is the sum of the absolute values of the changes from the planned program in each program examined. The score reported is the ratio of the absolute value sum to the sum of planned funding for FY 2000 through FY 2004. The median R&D funding instability score for the 15 programs for which data were

Table 4.1
Army Development Program Funding Instability, FY 2000–2004

Program	Amount (2004 $M)	Instability Score	Note
Longbow Apache	11.2	8	
Comanche	495.8	15	
ATIRCM/CMWS	33.6	22	
FAAD C2I	20.3	24	1999–2003
CH-47F	285.0	38	
FBCB2	87.1	41	
FMTV	4.3	42	
PAC-3 Fire Unit	15.2	45	Median
Longbow Hellfire	14.8	47	
Bradley Upgrade	38.3	49	1999–2003
Abrams Upgrade	31.9	58	
AN/TYQ-45 MCS	81.8	60	
Stryker	379.9	107	
Javelin	3.8	117	
GMLRS	143.9	212	
Average:	92.7	59	
Median:	31.9	45	

SOURCE: Calculated from *Selected Acquisition Report* data.

available was 45.[4] This is considerably higher than the median score of 21 that was the basis for the earlier study results. The scores for those 18 programs ranged from 2 to 86. The larger average score for Army R&D program funding instability reflects the high scores in the GMLRS, Javelin, and Stryker programs. Thus, we find that R&D funding instability was greater in 2000 through 2004 than in the 1980s and 1990s.

Procurement Programs

Table 4.2 shows the procurement funding instability results, which present a contrast to the results for R&D funding instability. We were able to calculate procurement instability scores for 13 of the 16 Army

Table 4.2
Army Procurement Program Funding Instability, FY 2000–2004

Program	Amount (2004 $M)	Instability Score	Note
Longbow Hellfire	13.0	1	
Longbow Apache	165.7	4	
FMTV	376.5	14	
Javelin	210.0	15	
FAAD C2I	45.9	17	1999–2003
Bradley Upgrade	345.9	18	
Abrams Upgrade	643.9	22	Median
CH-47F	169.1	22	
FBCB2	127.4	23	
AN/TYQ-45 MCS	60.4	37	
GMLRS	139.8	69	
ATIRCM/CMWS	183.9	73	
PAC-3 Fire Unit	257.3	153	
Average:	210.6	36	
Median:	169.1	22	

SOURCE: Calculated from *Selected Acquisition Report* data.

[4] No score could be computed for the PAC-3 missile program because it had no planned RDT&E spending for the period.

programs active in the FY 2000–2004 period.[5] For the 13, the median instability score for procurement programs was 22, which is much smaller than the median instability score of 65 recorded for the sample of 18 Army procurement programs from the 1980s and 1990s.

Comparison

Table 4.3 summarizes the results of this comparison of the earlier Army programs with the more recent ones. The data show that Army R&D program funding instability was greater in the more recent period than in the earlier periods (reported in Chapter Two). In sharp contrast, the data also show that funding instability in the more recent Army procurement programs was smaller than it had been earlier.

This pattern of change is consistent with the recent emphasis on transformation. During the 2000–2004 period, Army transformation activities focused for the most part on developing more-deployable, more-sustainable capabilities that would be procured as they came to fruition. Though some procurement programs were cut back

Table 4.3
Comparison of More-Recent and Earlier Funding Instability Measures for Army Programs

	2000–2004	Earlier
Development programs	15 programs	18 programs
Minimum	8	2
Median	59	26
Mean	45	21
Maximum	212	86
Procurement programs	13 programs	18 programs
Minimum	1	18
Median	35	61
Mean	18	56
Maximum	153	100

SOURCE: Calculated from *Selected Acquisition Report* data.

---

5 The three exceptions were the Comanche, Stryker, and PAC-3 missile programs, which did not have procurement plans in December 1999.

sharply (e.g., the Abrams Upgrade), many "legacy to future" programs continued with what were mostly no more than modest deviations from planned funding levels.

How Does the Army's More-Recent Funding Instability Compare with That of the Air Force?

This section describes how we compared the Army's FY 2000–2004 funding instability with that of the Air Force. The measures of instability for Army programs have been described. Our Air Force measures were based on Air Force data from SARs for the same period used for the Army (i.e., December 1999 and December 2003), and were calculated using the techniques described and used earlier. In these comparisons, the fiscal and military planning environment was the same for both services. Both were pressing transformations and adapting to the requirements of the Global War on Terrorism. Budgets were growing, but Army RDT&E funding was going up at a faster rate than were the Air Force's RDT&E accounts. In contrast, Air Force procurement was growing more rapidly than Army procurement between 2000 and 2004.[6]

In general, the Air Force's planned RDT&E and procurement spending amounts were significantly greater than the Army's. For the five-year period examined, the Air Force's average planned development program ($510 million) was 60 percent larger than the Army's ($318 million). The Air Force and Army procurement programs were both much larger than their RDT&E programs; but here, too, the average Air Force program was larger—in this case, 67 percent larger—than the average Army. Since our funding instability measures are expressed relative to the size of the amounts planned, such differences in the size of the planned amount did not affect the comparisons.

---

[6] Since both services' budgets for FY 2003 and FY 2004 contain unknown amounts of direct wartime acquisition costs that cannot be easily separated, these comparisons may overestimate the growth in their investment spending.

Air Force Research and Development Programs

The sample of 19 Air Force RDT&E programs included three for which the planned RDT&E funding was zero and for which we therefore could not compute a funding instability score. However, each program experienced instability, because funding was added where none had been planned. The planned Air Force development programs were large. In particular, the F-22 Raptor program was slated to expend $4,747 million in the five years from 2000 through 2004.[7] This amount alone is almost identical to the amount of funding that was planned for all 15 of the selected Army programs combined.

Table 4.4 shows the results for the 16 Air Force development programs for which we could calculate funding instability. As can be seen, these 16 had a median funding instability score of 22, and the scores ranged from 7 to 851. The reason for the JDAM program earning the highest score was that its planned RDT&E funding was $9.6 million, but another $81.8 million was then added between 2000 and 2004. For SBIRS High, whose 71 made it the second-highest scorer, planned funding was at $1,772 million and was then increased by $1,221 million between 2000 and 2004.

Air Force Procurement Programs

Table 4.5 shows the results we achieved for Air Force procurement program funding instability using the same set of metrics we used for the R&D program results. As was the case for the development programs, the average size of the Air Force procurement programs was significantly greater than that of the Army procurement programs.

Two of the Air Force procurement programs, the F-22 Raptor and the C-17, accounted for 71 percent of planned spending, totaling $38,100 million. The Air Force's funding for procurement was more concentrated than the Army's. As can be seen in Table 4.5, the F-22's amount of funding instability resulted in a low score of just 4. And even though the C-17 program experienced funding instability of over $3.8 billion, its planned spending was $13.4 billion, so its procurement funding instability score was 28. The highest funding instability

---

7 This and all other amounts in this discussion are in FY 2004 dollars.

Table 4.4
Air Force Development Program Funding Instability, FY 2000–2004

Program	Amount (2004 $M)	Instability Score	Note
Advanced Medium Range Air-to-Air Missile (AMRAAM)	16.0	7	
Joint Primary Aircraft Training System (JPATS)	5.1	8	
National Polar-Orbiting Operational Environmental Satellite System (NPOESS)	156.1	9	
Joint Air-to-Surface Standoff Missile (JASSM)	56.9	14	
B-1B Computer Upgrade (CMUP)	14.6	14	
Evolved Expendable Launch Vehicle (EELV)	182.8	19	
Global Broadcast Service (GBS)	33.9	20	
C-17A Globemaster III	163.3	22	Median
NAVSTAR (space system)	214.3	25	
National Airspace System (NAS)	0.8	29	
Joint Surveillance Target Attack Radar System (JSTARS)	128.3	35	
Raptor (F/A-22)	1,756.6	37	
Minuteman III Propulsion Replacement Program (MMIII PRP)	11.9	38	
NAVSTAR (user equipment)	255.3	48	
Space Based Infrared System (SBIRS) High	1,266.5	71	
Joint Direct Attack Munition (JDAM)	81.8	851	
Average:	284.2	26	
Median:	128.0	22	

SOURCE: Calculated from *Selected Acquisition Report* data.

score was the one associated with the Air Force's smallest planned procurement program.

Comparison

Table 4.6 compares key measures for Army and Air Force development and procurement funding instability. As can be seen, except for JDAM's extremely large RDT&E funding instability, the Army and Air Force measures are generally comparable. The RDT&E median measure for the Army, at 45, is, however, notably larger than that for the Air Force. But the Army's average, or mean, score for development

Table 4.5
Air Force Procurement Program Funding Instability, FY 2000–2004

Program	Amount (2004 $M)	Instability Score	Note
JASSM	1.7	1	
F/A-22	509.5	4	
AMRAAM	48.4	9	
MMIII PRP	123.2	13	
JPATS	159.3	18	
B-1B CMUP	26.1	19	
NAS	52.5	19	
NAVSTAR (space system)	248.3	21	
C-17A	3,817.9	28	Median
MMIII Guidance Replacement System (GRP)	265.1	28	
Airborne Warning and Control System (AWACS)	117.9	38	
EELV	991.1	52	
Hercules (C-130J)	638.7	60	
JDAM	660.1	69	
NAVSTAR (user equipment)	495.8	84	
JSTARS	714.4	88	
SBIRS High	547.9	121	
GBS	43.9	156	
Average:	525.7	46	
Median:	256.7	28	

SOURCE: Calculated from recent *Selected Acquisition Report* data.

program funding instability is less than the Air Force's. This is largely because the Air Force average is heavily influenced by the JDAM program's extremely large funding instability.

The analysis of Air Force procurement programs yielded results similar to those for the Army. We were able to compute a procurement funding instability score for 18 of the 19 Air Force programs. For those 18, the median funding instability score was 28, and the scores ranged from 1 to 156. These results are very much in line with those for the Army, which had a median instability score of 22 and scores that ranged from 1 to 153.

Table 4.6
Measures of Army and Air Force Program Funding Instability, FY 2000–2004

	Army	Air Force
Development programs		
Minimum	8	7
Median	45	22
Mean	59	80
Maximum	212	851
Procurement programs		
Minimum	1	1
Median	18	28
Mean	35	44
Maximum	153	156

SOURCE: Calculated from *Selected Acquisition Report* data.

The results in Table 4.6 indicate that both the median and the mean of funding instability were lower for the Army procurement programs than for the Air Force procurement programs. This does not support the hypothesis that the Army experiences greater procurement funding instability in its programs than the Air Force does.

Our second analysis of funding instability in the Air Force and Army programs compared planned and actual program spending to determine the frequency of funding reductions and increases. In this analysis, there were five observations for each program, one for each year of planned funding. Each time planned funding for any year was compared with actual funding, the result could be no change, an increase, or a decrease.[8] For example, 13 Army RDT&E programs allowed us to make 65 comparisons of planned and actual funding. Of these 65 comparisons, 14.7 percent showed that funding was as originally planned, 44.0 percent showed that there had been funding increases, and 41.3 percent showed that there had been funding cuts. The same sorts of calculations were made for Army procurement

---

8 The smallest change counted here is $0.1 million, or $100,000. Over the four databases examined in this analysis, exactly 10 percent were less that $1.0 million (not including "no change" observations). In both services, changes that were less than 10 percent of the planned funding amounted to approximately 20 percent of total observations.

programs and for Air Force RDT&E and procurement programs. The results for all four calculations are summarized in Table 4.7.

Overall, the results again demonstrate that funding change patterns within the Army and Air Force were similar. In both services' RDT&E and procurement programs, the most frequent funding change was an increase above the planned level. But funding cuts were almost equally as frequent.

The net results of the changes (shown at the bottom of Table 4.7) were generally to increase total FY 2000–2004 funding for the programs examined, the amounts of the increases ranging from 14 to 26 percent. However, the procurement funding of the Army programs examined was reduced by 4 percent.

What the Analysis of More-Recent Experience Reveals

The extent of R&D funding instability experienced by Army programs from FY 2000 through FY 2004 was considerably greater than what Army programs experienced in the 1980s and 1990s. The higher degree of instability may well be the result of Army leaders' efforts to create

Table 4.7
Direction and Results of Individual-Year Program Changes

	Percentage of Comparisons	
	Army	**Air Force**
RDT&E direction		
No change	14.7	7.5
Increase	44.0	48.7
Cut	41.3	43.8
Procurement direction		
No change	3.0	4.4
Increase	49.2	51.1
Cut	47.7	44.4
Net result for program funding		
RDT&E	+20	+26
Procurement	–4	+14

SOURCE: Calculated from *Selected Acquisition Report* data.

new transformational programs with funds made available through modification or cancellation of ongoing programs. Army programs also experienced much greater instability in their R&D funding in the FY 2000–2004 period than did Air Force programs.

In marked contrast to the R&D results, Army and Air Force programs experienced roughly the same degree of procurement instability in the FY 2000–2004 period. And the Army's procurement funding instability in this period was considerably less than what the Army had been experienced in the 1980s and 1990s. One possible conclusion is that the set of programs that had passed Milestone C and entered their production phase by the year 2000 were in consonance with Army leaders' intermediate goals to maintain and modernize the force at the same time they were pursuing transformation with newer systems. Change was the norm in both services. Each development program and each procurement program studied experienced at least one change in funding, and for the majority of programs, actual funding differed from planned spending in every year.

As the bottom portion of Table 4.7 shows, the net result of the funding changes was generally higher funding. Overall, funding for Army development programs was increased by 20 percent. Similarly, the Air Force increased its planned development program spending by 14 percent. In procurement programs, the Air Force added nearly $5.3 billion, or 26 percent, to its planned funding. However, the Army cut about $636 million, or 4 percent, of its planned procurement spending for FY 2000 through FY 2004.

Funding deviates from plans for many reasons, as was suggested by our case studies. Without examining each program in detail, there is no way to say why some programs experienced high funding instability while others did not. Nor is there evidence for equating high instability with mismanagement—large funding changes are often made for what appear to be valid reasons, such as changes in force structure requirements. Changes in R&D funding may reflect instances in which funds are applied to solve unexpected technical problems. Procurement changes typically arise from changes in acquisition quantities. Without further research, one cannot come to the conclusion that a higher level of funding instability—such as that observed for the Army, compared with the Air Force, R&D programs in 2000–2004—means that R&D outcomes for Army programs were adversely affected.

CHAPTER FIVE

Findings

This study was motivated by a concern that funding changes for Army programs had occurred without sufficient attention being paid to their effects on program management. In this view, Army program managers were burdened by unstable funding, which leads to adverse outcomes. Cost growth, schedule slippage, and performance compromises ultimately lead to reductions in number, timeliness, and capability when new systems are fielded. Some earlier RAND work suggested that the growth in development cost experienced in Army programs did not differ significantly from that experienced in the other services. However, both procurement cost growth and schedule slippage were, overall, greater in Army programs.

For our analysis, we broke the problem into several parts. We first developed a metric for funding instability and used it to assess funding instability in Army R&D programs in the first years following achievement of Milestone B. We also assessed funding instability for procurement programs, in this case for the first five years of planned procurement after Milestone B. The amount of relevant and consistent data that was available limited our analysis to at most 18 programs.

Our results showed that all of the 18 Army programs examined had faced funding instability, but that the amount of instability had varied widely. In general, funding instability was smaller for development programs than for procurement programs. Our development funding instability scores ranged from a minimum of <1 to a high of 86; our procurement funding instability ranged from 18 to just under 100. These results suggest that funding instability is the norm. However, measuring instability is only the first step, because instability scores tell nothing about the relationship between funding instability and adverse outcomes.

Our next step was to see whether funding instability was associated with adverse outcomes in Army programs. Metrics were developed for development cost growth, procurement cost growth, and schedule slippage for each of the 18 Army programs selected for analysis. Both casual inspection of the data and statistical analysis confirmed that funding instability was generally not associated with adverse outcomes. Funding instability in Army development programs was not associated with development cost growth, procurement cost growth, or schedule slippage. Funding instability in procurement programs was not associated with procurement cost growth. We did find a significant association between procurement funding instability and schedule slippage. However, one cannot come to conclusions about cause and effect based on a measure of association: Funding instability may lead to slipped schedules, or slipped schedules may lead to funding instability. A more refined level of analysis is required to determine cause and effect.

We undertook three case studies to allow a closer look at the causes and effects of funding instability. Since we were seeking the effects of funding instability, the case studies were chosen from Army programs that had experienced the largest funding instability. Our approach to the case studies included recognizing that funding instability can arise from root causes external to the Army, as well as from decisions (whether good or mistaken) made within the Army. Our study team closely examined available records of program management and interviewed participants who had been in the program when the funding instability occurred.

The results of the case studies did not point to changes made by Army leadership as a major source of funding instability. Examination of the Javelin program found both internal and external sources for funding instability. The Javelin development program's high funding instability arose from ambitious technical goals that had been set within the Army. These goals, which in hindsight might be considered a mistake, were reviewed and approved by OSD at Milestone B. The Javelin procurement program's equally high funding instability arose as a consequence of the Cold War's demise and the ensuing reduction in requirements as Army force structure was cut. Though the latter changes were implemented within the Army, they were a direct result of new strategic and funding guidance from DoD leadership. These changes caused a two-year stretch of the Javelin development schedule and Nunn-McCurdy program breaches.

The FAAD C2I case study yielded a similar result. An ambitious program relying on off-the-shelf items was frustrated when GFE was not delivered as planned. Then an ill-considered request for proposals failed to draw an acceptable response from industry. As a consequence, essentially none of the originally planned procurement was accomplished as planned, even though program priorities were restructured and some technical requirements relaxed. The FAAD C2I program was also hit by the force structure reductions following the end of the Cold War. Requirements for fielded FAAD C2I were cut dramatically.

The third case study, that of the CH-47 program, reflects more-recent experience than the other two case studies do (it was approved for development in 1998). External changes—unanticipated increases in the contractor's costs—led the program manager to recommend slipping the start of low-rate production and extending development by 12 months to cover the gap. The events in the Global War on Terrorism led to a decision to reorder planned production by moving the SOF version of the helicopter to the front of the production schedule. Because this is a recent case, it is too early to assess the effects of the resulting funding instability.

It is likely that Army leaders' concerns about the effects of paying too little attention to how funding instability impacts program management were grounded in more-recent experience. We thus decided to expand the analysis to evaluate funding instability in the 2000–2004 period as the third part of our investigation. This necessarily involved a somewhat different set of Army development and procurement programs, since some had been completed, some canceled, and some added.

Given the many changes in the defense environment since 2000 (a new administration, increased emphasis on transformation, the 9/11 attacks, and the Global War on Terrorism), a higher level of funding instability might be expected. We found that funding instability in Army R&D programs had indeed increased. However, we also found that funding instability in Army procurement had fallen.

The data for 2000 through 2004 allowed us to compare the Army's more-recent funding instability with that in 19 Air Force programs. Our analysis of funding instability in the Air Force programs used the same data source (SARs) and the same metrics that we had used in analyzing Army programs. As judged by the median level of funding instability, the Army had a higher level of funding instability than the Air Force in its development programs, but a lower funding instability in its procurement programs. Examination of the individual annual deviations from planned funding revealed similar patterns in the Army and Air Force.

In sum, our analysis of funding instability in the Army found wide variation across programs but failed to show that funding instability was associated with cost growth. However, procurement program funding instability and schedule slippage were associated. And changes to planned yearly funding were the norm. Our three case studies suggest that most funding instability arose from understandable and important external root causes, such as the end of the Cold War and the Global War on Terrorism. In two of the three study cases, overly ambitious development programs subsequently experienced development funding instability. Our comparison of more-recent funding instability experiences in Army and Air force programs did not reveal significant differences between the two services. We discovered no reason in our analyses to believe that Army program managers faced unusually large or inexplicable funding changes that led to adverse outcomes.

APPENDIX A

Army Program Descriptions

Abrams Upgrade

The Abrams tank closes with and destroys enemy forces on the integrated battlefield using mobility, firepower, situational awareness, and shock effect. The combination of the 120mm main gun with the 1,500-hp turbine engine and special armor makes the Abrams tank particularly suitable for attacking or defending against large concentrations of heavy armor forces in a highly lethal battlefield.

The Abrams Upgrade—designated the M1A2—provides the Abrams tank with the necessary improvements in lethality, survivability, and ability to engage targets required to defeat advanced threats. The M1A2 includes a commander's independent thermal viewer, an improved commander's weapon station, position navigation equipment, a distributed data and power architecture, an embedded diagnostic system, and an improved fire control system. The M1A2 System Enhancement Program adds second-generation thermal sensors, a Thermal Management System, and upgrades to processors/memory so that the M1A2 can use the Army's common C2 software, thereby enabling the rapid transfer of digital situational data and overlays.

Advanced Threat Infrared Countermeasure/Common Missile Warning System (ATIRCM/CMWS)

The ATIRCM/CMWS is a program designed to develop, test, and integrate defensive infrared countermeasure capabilities into existing current-generation host platforms for more-effective protection against a greater number of infrared guided missile threats. The operational requirements concept for infrared countermeasure systems is known as the Suite of Integrated Infrared Countermeasures, an integrated warning and countermeasure system for enhancing aircraft survivability against infrared guided missile threats. The core element of the Suite is the ATIRCM/CMWS. A subsystem to a host aircraft, the ATIRCM/CMWS is an integrated ultraviolet missile warning system and an infrared Lamp/Laser Jamming and Improved Countermeasure Dispenser. It also functions as a stand-alone system with the capability to detect missiles and provide audible and visual warnings to pilots. When installed with the Improved Countermeasure Dispenser, the ATIRCM/CMWS activates expendables to provide some protection. The ATIRCM/CMWS is the key infrared survivability system for Future Force Army aircraft.

AN/TYQ-45 Maneuver Control System (AN/TYQ-45 MCS)

The AN/TYQ-45 MCS is a C2 system that provides commanders and staff, at corps through battalion, with up-to-date information for quick decisions and efficient use of firepower and maneuver resources. The AN/TYQ-45 MCS database provides decision support information and functional tools in both text and map graphics form. This system also automates the preparation and distribution of operations orders and reports to facilitate execution of the commander's decisions. Reports received through the MCS automatically update the database, ensuring current tactical information. The MCS minimizes life-cycle costs and capitalizes on state-of-the-art ruggedized commercial equipment through use of Common Hardware/Systems computers and peripheral hardware. The AN/TYQ-45 MCS also uses ruggedized commercial notebook computers to enhance software development, support, and training.

Black Hawk Utility Helicopter (UH-60A/L)

The UH-60A/L is a twin-engine Army utility helicopter that is used in air assault, air cavalry, and aeromedical evacuation missions. It is sized as an infantry squad assault helicopter capable of carrying up to 14 troops. It transports troops and equipment to combat, resupplies troops in combat, and performs transport functions associated with aeromedical evacuation, repositioning of reserves, and C2.

Bradley Fighting Vehicle System Upgrade

The Bradley Upgrade provides infantry and cavalry fighting vehicles with digital C2 capabilities, increased situational awareness, enhanced lethality, increased survivability, and improved sustainability and supportability. This upgrade consists of two second-generation forward-looking infrared sensors in the Improved Bradley Acquisition System and Commander's Independent Sight that together provide "hunter-killer target handoff" capability with: a ballistic fire control system; embedded diagnostics; an integrated combat continuous-wave digital communications suite hosting a Force XXI Battle Command Brigade-and-Below package with digital maps, messages, and friend/foe situational awareness; a position navigation system with a global positioning system (GPS) and inertial navigation system; and enhanced squad situational awareness with a squad leader display integrated into vehicle digital images and integrated combat C2.

Brilliant Anti-Armor Technology Preplanned Product Improvement (BAT P3I)

The Brilliant Anti-Armor Technology (BAT) submunition provides deep fires to Army Objective Force and Joint forces commanders to delay and disrupt enemy armored forces at ranges in excess of 100 kilometers. BAT is a top attack submunition with acoustic and infrared seekers working in tandem for autonomous attack of moving armor.

The preplanned product improvement program for BAT adds cold stationary armor, heavy multiple rocket launchers, and surface-to-surface missile transporter erector launchers to the target set through seeker and warhead improvements. BAT Preplanned Product Improvement (P3I) submunitions are carried deep into enemy territory by the Army Tactical Missile System Block II missile and then dispensed to attack and destroy targets.

Chinook Improved Cargo Helicopter (CH-47F)

The CH-47F program is a remanufacturing program that will extend the service life of the current heavy-lift helicopter fleet by 20 years. The CH-47F aircraft incorporates improvements to airframe reliability and maintainability through stiffened structural components to reduce vibration; provides an avionics architecture compliant with the applicable information technology standards contained in the DoD Joint Technical Architecture, Joint Technical Architecture–Army, and is interoperable with defense systems through a digital communication/navigation cockpit upgrade; increases lift capability and range; and complies with Global Air Traffic Management, Air Warrior, and Digital Source Collector requirements. The CH-47F transports ground forces, supplies, and other battle-critical cargo and is designed to be strategically responsive across the full spectrum of Army operations, including support, coverage, and sustainment of maneuver, fire support, air defense, and survivability missions.

Comanche (terminated)

The RAH-66 Comanche helicopter was intended to be the Army's next-generation armed reconnaissance aircraft system. It was to provide the Army with network-centric capability from a joint and a combined perspective. Comanche's technology represented a system capable of operating in adverse weather conditions across a wide spectrum of threat environments. Comanche's innovative design was expected to lower operating costs through the use of integrated diagnostics and component functional partitioning, eliminating the requirement for Aviation Intermediate Maintenance. Comanche's advanced airframe design incorporates composite airframe structures, a bearingless main rotor system, and reduced signatures. The Comanche Mission Equipment Package was to feature an open systems architecture integrating second-generation target acquisition and night vision sensors. Pilot workload from targeting to navigation is significantly reduced due to introduction of cognitive-decision-aiding and fully integrated weapon systems.

Combat Service Support Control Systems (CSSCS)

The CSSCS is an automated C2 system supporting the combat service support component of the Army Battle Command System. The CSSCS rapidly collects, processes, and distributes critical logistical, personnel, medical, and transportation information to assist commanders with planning and execution of combat service support and C2 operations. The CSSCS comprises Army Battle Command System common hardware, common operating environment software, and CSSCS-unique software. The CSSCS hardware and software are housed in the Standard Integrated Command Post System family of shelters.

Family of Medium Tactical Vehicles (FMTV)

FMTV is a complete series of trucks based on a common chassis, varied by payload and mission. The Light Medium Tactical Vehicle has a 2.5-ton capacity and consists of van and cargo models. The Medium Tactical Vehicle has a 5-ton capacity and consists of cargo, tractor, van, wrecker, tanker, and dump-truck models. Subvariants include the Air Drop, trailer airdrop, and a water tanker. More than 80 percent commonality of parts among the variants reduces operational and support costs. FMTV is designed to be rapidly deployable worldwide and to operate on primary roads, secondary roads, trails, and cross-country terrain in all climate conditions.

Force XXI Battle Command Brigade and Below (FBCB2)

The FBCB2 is a digital, battle command information system that provides integrated, on-the-move, timely, relevant battle command information technology to allow commanders to concentrate combat system effects rather than combat forces, enabling units to be both more survivable and more lethal. FBCB2 provides the capability to pass orders and graphics that allow the warfighter to visualize the commander's intent and scheme of maneuver. FBCB2 affords combat forces the capability to retain the tactical/operational initiatives under all mission, enemy, terrain, troops, and time-available conditions to enable faster decisions and real/near-real time communications and responses. The system includes a Pentium-based processor, display unit, keyboard, and removable hard-disk drive cartridge. FBCB2 supports situational awareness (Blue and Red force positions) and C2 down to the soldier/platform level across the Battlefield Operating Systems and echelons. As the key component of the Army Battle Command System (ABCS), FBCB2 completes the information flow process from brigade to platform and across platforms within the brigade task force and across brigade boundaries.

Forward Area Air Defense Command, Control, and Intelligence (FAAD C2I)

The FAAD C2I system is the air defense node of the Army Tactical Command and Control System. It provides short-range air-defense information to support the C2 decision process at various levels of command. It ties weapons together with a C2I network and integrates the FAAD System (FAADS) into the Army Battle Command System Architecture. The FAAD C2I system integrates weapons, sensors, communications, and C2I architecture to counter the entire spectrum of the air threat to the divisional forward area. This mission is accomplished through collection, digital processing, and dissemination of target information, air threat warning, and C2 information. The FAAD C2I system also provides target data processing and display capabilities at the Air Battle Management Operations Center, the Army Airspace Command and Control element, Sensor/Command and Control node, Battery, Platoon/Section, and Fire Unit levels.

Guided Multiple Launch Rocket System (GMLRS)

The mission of the GMLRS is to attack/neutralize/suppress/destroy targets using indirect precision fires. GMLRS provides field artillery units with medium- and long-range fires while supporting brigade, division, corps, theater, joint/coalition forces and Marine air-ground task forces in full, limited, or expeditionary operations. GMLRS uses an inertial measuring unit with GPS assistance to guide the rocket to a specific point to deliver effects on a target. GMLRS is fired from the Multiple Launch Rocket System M270A1 tracked launcher or from the light wheeled launcher, the High Mobility Artillery Rocket System. GMLRS is transported and fired in a rocket pod that consists of six rockets with a rocket pod container.

Javelin

The Javelin system is a medium-range, imaging infrared, fire-and-forget, man-portable, antitank weapon system. The Javelin tactical system is composed of a tactical round and a command launch unit (CLU). The missile is sealed in a disposable launch tube assembly; it comprises the seeker, guidance electronics, warhead and fuse, propulsion unit, and the control actuator system. The CLU consists of an integral visible day telescope and a long-wavelength infrared night sight with wide and narrow fields of view. The CLU is used for battlefield surveillance, target acquisition, missile launch, and damage assessment. The Javelin system may be used in the normal top attack mode or in direct mode for engaging targets under cover. Javelin is capable of defeating conventional and reactive armor in day and night engagements and has a soft launch capability that enables it to be fired from enclosures or covered fighting positions. The Javelin system replaces the Dragon system.

Joint Surveillance Target Attack Radar System (JSTARS)

JSTARS is a surveillance, battle management, and targeting radar system. The JSTARS radar is an airborne multimode radar system that is carried aboard a modified E-8 aircraft.

JSTARS Common Ground Station (JSTARS CGS)

The JSTARS CGS is a mobile C2 vehicle designed to receive targeting and surveillance information from the airborne JSTARS through an omni-directional radar link. JSTARS CGS also receives and processes intelligence data from unmanned aerial vehicles, Commander's Tactical Terminal, and Airborne Reconnaissance Low.

JSTARS Ground Station Module (JSTARS GSM)

The JSTARS GSM is a mobile, multisensor, imagery intelligence, tactical-data processing and evaluation system. The GSM is a subcomponent of JSTARS and is designed to detect, locate, and track moving and stationary ground equipment targets that are beyond the forward line of own troops. The GSM processes data from the JSTARS aircraft Commanders Tactical Terminals, Joint Tactical Terminal, and unmanned aerial vehicles. The GSM disseminates intelligence, battle management, and targeting data to Army command, control, communications, and intelligence nodes by local area network, wire, or radio communications, thus enabling integrated battle management, surveillance, targeting, and interdiction plans to be developed/executed using near-real-time data.

Longbow Apache Attack Helicopter

The Longbow Apache is the Army's premier attack helicopter.

Longbow Apache Airframe (AFM)

The Longbow Apache AFM program is the AH-64D aircraft itself (excluding the radar, which is a separately managed program). The AH-64D is an upgraded remanufactured version of the AH-64A; it incorporates updated T700-GE-701C engines and a fully integrated cockpit. In addition, it has improved survivability, communications, and navigation capabilities.

Longbow Apache Fire Control Radar (FCR)

The Longbow Apache FCR is a mast-mounted millimeter-wave target-acquisition system that is integrated into the AH-64 Apache attack helicopter airframe. The Longbow Apache FCR provides rapid automatic detection, classification, and prioritization of multiple ground and air targets in adverse weather and under battlefield obscurants. The FCR has four modes: (1) air targeting mode, which detects, classifies, and prioritizes fixed and rotary wing threats; (2) ground targeting mode, which detects, classifies, and prioritizes ground and air targets; (3) terrain profiling mode, which provides obstacle-detection and adverse-weather pilot aids to the Longbow crew; (4) built-in-test mode, which monitors radar performance in flight and isolates electronic failures before and during maintenance.

Longbow Hellfire

The Longbow Hellfire is an air-to-ground point-target, precision-strike missile system designed to defeat individual hard point targets. The Longbow Hellfire uses a radio-frequency guidance section that provides capabilities for lock-on before or after launch. This missile system is employed on the AH-64D Longbow Apache attack helicopter, providing it with the capability to engage targets both day and night in all weather conditions and with battlefield obscurants present. The Longbow Hellfire also provides a fire-and-forget capability against given target sets.

Patriot Advanced Capability 3 (PAC-3)

Patriot is a high-to-low-altitude, air-defense missile system that provides air defense of ground combat forces and high-value assets. Patriot is designed to cope with enemy defense suppression tactics that may include tactical ballistic missiles, cruise missiles, anti-radio missiles, and advanced aircraft employing saturation, maneuver, sophisticated electronic countermeasures (ECMs), and low radar cross-section. The system can conduct multiple simultaneous engagements of high-performance air-breathing targets and tactical ballistic missiles with a high probability of target kill. The Patriot system will provide air defense protection in all weather conditions and hostile ECM environments.

The PAC-3 program is the result of a series of integrated, phased system improvements fielded in combination with the PAC-3 missile. The PAC-3 missile is a high-velocity hit-to-kill, surface-to-air missile capable of intercepting and destroying tactical missiles and air-breathing threats. It provides the range, accuracy, and lethality to effectively defend against tactical missiles with conventional high explosive, biological, chemical, and nuclear warheads. The missile uses a solid-propellant rocket motor, aerodynamic vane controls, and inertial guidance to navigate to an intercept point. Shortly before arrival at the intercept point, the missile's rate of spin increases, the onboard radar homing seeker acquires the target, and terminal homing guidance is initiated to achieve hit-to-kill by high-resolution maneuvers.

Secure Mobile Anti-Jam Reliable Tactical Terminal (SMART-T)

The SMART-T provides range extension capability to the Army's Mobile Subscriber Equipment and Future Warfighter Information Network–Tactical. More specifically, it provides a satellite interface to permit uninterrupted voice and data communication as advancing forces move beyond the line-of-sight capability of terrestrial communications systems. The SMART-T supports echelons corps and below and special contingency operations. The SMART-T equipment communicates at both low and medium data rates and provides the security, mobility, and anti-jam capability required to defeat the threat. The SMART-T has inherent low probability of interception and low probability of detection to avoid being targeted for destruction, jamming, or eavesdropping. The prime mover is a High Mobility Multi-purpose Wheeled Vehicle that carries all electronics, power generation, and a self-erectable antenna. These terminals, which can also be used in a fixed configuration, increase the tactical utility of the Milstar System.

Sense and Destroy Armor Submunition (SADARM 155)

The SADARM 155 is a smart munition that provides enhanced counter-fire capability for the 155mm howitzer delivery system. It enables the howitzer to attack targets beyond the forward line of own troops in a fire-and-forget mode. This indirect fire mission can be accomplished day or night and under adverse weather conditions, degraded battlefield conditions, and in nuclear, biological, and chemical environments. Designed for use against self-propelled howitzers, lightly armored personnel carriers, and other stationary armored threat vehicles encountered in counter-fire, close support, suppression of enemy air defense, and interdiction, the SADARM 155 has five major parts: (1) multimode sensor with infrared and active-passive millimeter wave; (2) lethal mechanism with explosively formed penetrator; (3) parachutes that control deceleration, spin, and descent velocity; (4) fuzing, safe, and arm device; and (5) carrier hardware.

Stryker

The Stryker family comprises two variant vehicles, the Infantry Carrier Vehicle (ICV) and the Mobile Gun System (MGS). There are nine different configurations for the ICV: basic ICV, Reconnaissance Vehicle, Mortar Carrier, Commander's Vehicle, Fire Support Vehicle, Engineer Squad Vehicle, Medical Evacuation Vehicle, Anti-Tank Guided Missile Vehicle, and Nuclear, Biological, Chemical (NBC) Reconnaissance Vehicle. The Stryker family of vehicles is air transportable in a C-130 aircraft, is capable of immediate employment upon arrival in the area of operations, and maximizes commonality among variants.

APPENDIX B

Air Force Program Descriptions

Advanced Medium Range Air-to-Air Missile (AMRAAM)

The AMRAAM system is an active radar-guided intercept missile with inherent electronic-protection capabilities for air-to-air applications against massed penetration aircraft. It is designed to augment the AIM-7 Sparrow. The AMRAAM program provides for acquisition of the most advanced all-weather, all-environment medium-range air-to-air missile system in response to U.S. Air Force, U.S. Navy, NATO, and other allied operational requirements for the 1989–2007 period.

Airborne Warning and Control System (AWACS)

AWACS detects and tracks enemy and friendly aircraft and air vehicles over a large component of the theater airspace; it is mounted in an Air Force E-3 aircraft. The Radar System Improvement Program (RSIP) modification is to provide new and improved capabilities for the Airborne Warning and Control System radar. The AWACS RSIP provides improvements in radar sensitivity/electronic counter countermeasures (ECCM) performance, radar performance monitoring and control, and reliability/maintainability (R&M) to maintain system effectiveness against the projected operational environment of the 1990s and into the next century.

B-1B Computer Upgrade (B-1B CMUP)

The B-1B CMUP consists of three efforts: replacing the mission computers themselves, replacing the defense system software suite, and adding a Joint Direct Attack Munition (JDAM) capability.

Computer Upgrade

The computer upgrade element of the B-1B CMUP is the major element of CMUP Block E. The program will replace six existing computers (controls and displays, guidance and navigation, weapon delivery, critical resources function, and terrain following) with four new computers. The current data transfer system will be replaced with a new one, and the avionics flight software will be converted from JOVIAL to Ada. The objective is to increase memory capacity, throughput, input/output, bandwidth, and growth potential; to improve reliability and maintainability; and to provide a flexible weapon capability. As a result, the B-1B will be able to carry and deliver three different types of weapons on the same sortie with a single software load.

Defensive Systems

The defensive systems upgrade will remove most of the ALQ-161 system and replace it with an AN/ALR-56M radar warning receiver and the Radio Frequency Countermeasures portion of the Navy's Integrated Defense Electronic Counter Measures (IDECM) program, which includes a techniques generator and a fiber-optic towed decoy. A new low-band onboard jammer will be installed to provide the requisite threat coverage. These new systems will significantly improve situational awareness and the B-1B's survivability in the medium- and high-altitude regimes, where most conventional missions will be conducted.

JDAM Capability

The Air Force has established a requirement to enhance the capability of the B-1B Lancer to perform near-precision attacks against all but heavily defended targets deep in enemy airspace during conventional operations. The requirement is satisfied with a material solution to provide the B-1B with improved lethality through integration of such near-precision conventional weapons as the JDAM. Implementation of MIL-STD-1760 electrical interconnect system, communication upgrades, and the GPS are included as part of the advanced munitions integration.

C-17A Globemaster III

The C-17A Globemaster III is a multi-engine, turbofan, wide-body, strategic airlift aircraft that enables the Air Force to rapidly project, reinforce, and sustain combat forces worldwide. This aircraft augments the C-5 and C-141 in inter-theater deployment, and the C-130 with intra-theater operations. The C-17 is capable of carrying outsized cargo over inter-theater ranges into austere airfields and introduces a direct deployment capability that significantly improves airlift responsiveness.

Significant features of the C-17 include: super-critical wing design and winglets to reduce drag and increase fuel efficiency and range; in-flight refueling capability to increase range; externally blown flap configuration, direct lift control spoilers, and high-impact landing gear system to enable the aircraft to operate into and out of small austere airfields; forward- and upward-directed thrust-reverser system to provide backup capability, reduce aircraft ramp space requirements, and minimize interference of dust and debris with the activities of ground personnel; cargo-door, ramp-airdrop, and cargo-restraint systems operable by a single loadmaster to permit immediate equipment offload without special handling equipment; two-person cockpit, with multifunction displays, to reduce complexity and improve reliability; built-in-test features to reduce maintenance and troubleshooting times; and walk-in avionics bays to improve accessibility. This aircraft was designed to have lower maintenance man-hours per flight hour than its predecessors.

Evolved Expendable Launch Vehicle (EELV)

The mission of the EELV is to partner with industry to develop a national launch capability that satisfies the government's National Mission Model requirements and reduces the cost of space launch by at least 25 percent over existing systems. The EELV system includes launch vehicles, infrastructure, support systems, and payload interfaces. EELV is a family of launch vehicles evolved from existing expendable launch systems or components thereof. EELV is supporting military, intelligence, and civil mission requirements (previously serviced by Titan II, Delta II, Atlas II, and Titan IV launch vehicles) in the National Mission Model through 2020.

Global Broadcast Service (GBS)

The GBS provides worldwide, high-capacity, one-way transmission of video, imagery, and other large data files in support of joint military forces in garrison, in transit, and in theater using satellite technology. GBS augments existing military satellite communication systems. Using wireless GBS satellite receiver systems, military users afloat and ashore receive live and recorded video information, large data files (such as weather maps and high-resolution imagery), and Internet-like services to perform their missions while retaining the mobility afforded by satellite-based communication.

Hercules (C-130J)

The C-130J is a medium-range, tactical airlift aircraft designed primarily for transporting cargo and personnel within a theater of operations. Variants of the C-130J perform other missions, including rescue and recovery, air refueling, special operations, fire fighting, and weather reconnaissance.

The C-130J can carry more than 40,000 pounds of cargo or be configured to carry up to 84 paratroopers. The enhanced cargo-handling system reduces crew workload and can be quickly adapted to accommodate any combination of passenger, cargo, or aeromedical airlift mission. Two primary methods of aerial delivery are used for equipment delivery: parachutes pulling the load from the aircraft, and the Container Delivery System that uses the force of gravity to pull supplies from the aircraft. The C-130J can also operate from austere landing zones with as little as 3,000 feet of dirt runway.

Joint Air-to-Surface Standoff Missile (JASSM)

JASSM is a next-generation air-to-surface missile that will enable Air Force and Navy bombers and fighters to destroy the enemy's war-sustaining capabilities from outside the ranges of enemy air defenses. The autonomous precision strike weapon will attack targets that are either fixed or relocatable, ranging from non-hardened above-ground targets to moderately hardened buried point targets. The system was designed to operate worldwide and have low operational support costs. The increased stand-off range of the JASSM Extended Range (JASSM-ER) will allow precision attack of targets, deeper into enemy territory, while minimizing the threat to the launch aircraft. JASSM does not replace any existing weapon system.

Joint Direct Attack Munition (JDAM)

JDAM is a joint Air Force/Navy program with the Air Force as the lead service. This program upgrades the existing inventory of general-purpose bombs by integrating the bombs with a guidance kit consisting of a GPS-aided inertial navigation system. JDAM provides an accurate, adverse-weather capability against mobile hard, mobile soft, fixed hard, fixed soft, and maritime targets.

Joint Primary Aircraft Training System (JPATS)

JPATS is an Air Force/Navy program to replace the T-37B and T-34C aircraft and the associated ground-based training systems, which are being used to train entry-level students in the fundamentals of flying so that these students can transition to advanced training tracks leading to qualification as military pilots, navigators, and naval flight officers.

The program represents a systems approach to aviator training that requires the purchase of air vehicles (782 production units), aircrew training devices (122), associated ground-based training devices, an integrated training management system, instructional courseware, and contractor logistics support. The Air Force will train at six bases, the Navy at three. Each operational training location will be equipped with a full complement of operational flight trainers, instrument flight trainers, unit training devices, and egress training devices. Courseware has been developed for the T-6A and converted from existing courseware for other platforms where appropriate. The Training Integrated Management System (TIMS) will provide a training and scheduling capability to tie together the efforts and activities of all Air Education and Training Command and Chief of Naval Air Training operating locations.

Joint Surveillance Target Attack Radar System (JSTARS)

JSTARS is a joint Army and Air Force program, with the Air Force as the lead service. The JSTARS system provides real-time wide-area surveillance of the battlefield and rear echelons. JSTARS is unique because it detects and tracks enemy armor, vehicles, and troops over a wide area in real-time using moving target indicator (MTI) and synthetic aperture radar techniques. JSTARS also plays a critical C2 battle-management role by providing precise real-time targeting information to direct attack aircraft, friendly artillery, and stand-off missile batteries. JSTARS can give the joint force commander a near-real-time look at enemy first and second echelon force buildups, force movements, and the enemy scheme-of-maneuver on the battlefield. It provides MTI/

synthetic aperture radar coverage of ground activity, with target identification and intelligence support from RIVET JOINT, and it works in concert with AWACS to provide a collaborative situation-awareness, battle-management, and precision-engagement capability for the joint force commander.

Minuteman III Guidance Replacement Program (MMIII GRP)

The MMIII GRP upgrades and extends the life of the Minuteman III guidance system through 2020. As a result of the recent Nuclear Posture Review, Minuteman III is projected to become the only land-based intercontinental ballistic missile in the U.S. nuclear arsenal when Peacekeeper is retired. The guidance electronics need to be replaced because the current electronic components continue to degrade and are becoming unreliable and unsupportable. The GRP replaces the 1960s guidance system electronics and protects the option for future implementation of the Mark 21 RV/W87 warhead and an advanced inertial measurement unit, if required.

Minuteman III Propulsion Replacement Program (MMIII PRP)

The MMIII PRP extends the life, maintains the performance, and improves the reliability of the Minuteman operational force by replacing the current fielded motors prior to the onset of deterioration or failure because of aging. The solid-propulsion systems now in the force began aging out in 2002 and must be replaced to support current force planning. The PRP reuses existing components to the greatest extent possible. During remanufacture, the solid rocket motors and interstage hardware and ordnance are being recycled from the force and remanufactured at a rate up to eight boosters per month during the FY 2000–2009 period.

National Airspace System (NAS)

The DoD NAS is composed of three subsystems: the Digital Airport Surveillance Radar (DASR), DoD Advanced Automation System (DAAS), and Voice Communication Switching System (VCSS).

The DoD NAS program will modernize the DoD radar approach-control facilities in parallel with the Federal Aviation Agency (FAA). This program provides systems and facilities compatible/interoperable with the FAA modernization, prevents DoD flight delays and cancellations, continues DoD's access into Special Use Airspace, provides transparent services to military and civil aircraft, replaces aging DoD air traffic control systems, and increases flight safety. NAS is a non-developmental item acquisition. DoD will upgrade voice, data, and sensor systems, as well as facility configurations and operations concepts. The NAS program also includes the Military Airspace Management System, an automated scheduling and utilization-reporting tool for special-use airspace that will enable DoD to more efficiently manage special-use airspace.

National Polar-Orbiting Operational Environmental Satellite System (NPOESS)

The NPOESS program is required to provide, for a period of at least ten years, a remote sensing capability to acquire, receive at ground terminals, and disseminate to processing centers global and regional environmental imagery and specialized meteorological, climatic, terrestrial, oceanographic, solar-geophysical, and other data supporting Department of Commerce/National Oceanic and Atmospheric Administration mission requirements, and DoD peacetime and wartime missions.

NAVSTAR Global Positioning System (NAVSTAR GPS)

The NAVSTAR GPS is a space-based radio positioning, navigation, and time-distribution system that provides precise, continuous, all-weather, common-grid positioning, velocity, navigation, and time-reference capability to civil, commercial, and military users worldwide. The military mission areas supported by this system include navigation and position fixing; air interdiction; close air support; special operations; strategic attack; counter-air and aerospace defense; theater and tactical command, control, communications and intelligence; precision munitions guidance; and ground/sea warfare. GPS also carries a suite of nuclear-detonation detection-system sensors as a secondary payload. These sensors provide worldwide, near-real time, three-dimensional location of nuclear detonations.

Raptor (F/A-22)

The F/A-22 is a multirole fighter designed to penetrate enemy airspace and achieve a first-look, first-shot, first-kill capability against multiple targets. The engineering and manufacturing development (EMD) program consists of design, fabrication, and development testing of nine EMD flight-test vehicles and 25 engines; update of the avionics flying test-bed and use of it to develop and integrate the EMD avionics suite; and design and development of the F/A-22 support and training system. The ongoing production program will deliver F/A-22s, along with the required alternate mission equipment, support equipment, and training systems. The F/A-22 is characterized by a low-observable, highly maneuverable airframe, engines capable of supersonic cruise, and advanced integrated avionics.

Space Based Infrared System (SBIRS) High

SBIRS is an integrated system consisting of multiple space and ground elements, with incremental deployment phasing, simultaneously satisfying requirements in the following mission areas: missile warning, missile defense, technical intelligence, and battle space characterization. The baseline architecture for SBIRS includes space elements in highly elliptical orbits and geosynchronous earth orbits, in addition to the following ground elements: a CONUS-based mission control station and mission control station backup, overseas relay ground stations, a multimission mobile processor, and associated communication links. The High component of the SBIR consists of five satellites (four operational and one spare) in geosynchronous orbits, two hosted sensors in highly elliptical orbits (platforms provided by another organization), and associated ground elements.

References

Birkler, John, John C. Graser, Mark V. Arena, Cynthia R. Cook, Gordon T. Lee, Mark A. Lorell, Giles K. Smith, Fred Timson, Obaid Younossi, and Jon Grossman, *Assessing Competitive Strategies for the Joint Strike Fighter: Opportunities and Options*, Santa Monica, Calif.: RAND Corporation, MR-1362-OSD/JSF, 2001. As of April 10, 2007:
http://www.rand.org/pubs/monograph_reports/MR1362/

Center for Naval Analysis, *The Army Acquisition Management Study: Congressional Mandate for Change*, Alexandria, Va.: Center for Naval Analysis, CRM D0003618.A2/Final, April 2001, p. 5.

Clark, Colin, "DoD Elevates Cost Focus," *Defense News*, 6 September 1999, p. 1.

Cohen, W. S., *Secretary of Defense 1998 Annual Report to the President and the Congress*, Washington, D.C., 1998.

Congressional Budget Office, *Cost Growth in Weapon Systems: Recent Experience and Possible Remedies*, Washington, D.C., October 1982.

———, *Effects of Weapon Procurement Stretch-outs on Costs and Schedules*, Washington, D.C., 1987.

DoD Directive 5000.1, "The Defense Acquisition System," May 12, 2003. As of May 14, 2007:
www.at.dod.mil/Docs/d50001p.pdf

Drezner, Jeffrey A., and Richard A. Krop, *The Use of Baselining in Acquisition Program Management*, Santa Monica, Calif.: RAND Corporation, MR-876-OSD, 1997. As of April 10, 2007:
http://www.rand.org/pubs/monograph_reports/MR876/

Drezner, Jeffrey A., and Giles K. Smith, *An Analysis of Weapon System Acquisition Schedules*, Santa Monica, Calif.: RAND Corporation, R-3937-ACQ, 1990. As of April 10, 2007:
http://www.rand.org/pubs/reports/R3937/

Drezner, Jeffrey A., Jeanne M. Jarvaise, Ron Hess, Daniel M. Norton, and Paul G. Hough, *An Analysis of Weapons System Cost Growth*, Santa Monica, Calif.: RAND

Corporation, MR-291-AF, 1993. As of April 10, 2007:
http://www.rand.org/pubs/monograph_reports/MR291/

Hess, Ron, *Review of Cost Improvement Literature with Emphasis on Synthetic Fuel Facilities and the Petroleum and Chemical Process Industries*, Santa Monica, Calif.: RAND Corporation, N-2273-SFC, 1985. As of April 10, 2007:
http://www.rand.org/pubs/notes/N2273/

Higbee, John, "Program Success Probability," Briefing, Fort Belvoir, Va.: Defense Acquisition University, June 2, 2004.

Hildebrand, Greg, Rob Leonard, and Jack Graser, "Joint Programs in the DoD: How Do They Compare with Single Service Programs? A Preliminary Look," Briefing, Santa Monica, Calif.: RAND Corporation, September 9, 2003.

Hough, Paul G., *Pitfalls in Calculating Cost Growth from Selected Acquisition Reports*, Santa Monica, Calif.: RAND Corporation, N-3136, 1992. As of April 10, 2007:
http://www.rand.org/pubs/notes/N3136/

Jaynes, Howard R., Jr., "Correlation Analysis: Army Acquisition Program Cycle Time and Cost Variation," Thesis, Monterey, Calif.: Naval Postgraduate School, 1999.

Leach, Bob, Office of the Under Secretary of Defense (Acquisition, Technology, and Logistics), *Acquisition Reporting Overview*, Presentation at the Defense Acquisition University Business Managers Conference, June 12–13, 2002.

McNicol, David L., "Growth in the Costs of Major Weapon Procurement Programs," Alexandria, Va.: Institute for Defense Analysis, IDA Paper P-3832, 2004.

Merrow, Edward W., *An Analysis of Cost Improvement in Chemical Process Technologies*, Santa Monica, Calif.: RAND Corporation, R-3357-DOD, 1989. As of April 10, 2007:
http://www.rand.org/pubs/reports/R3357/

Merrow, Edward W., Stephen E. Chapel, and Christopher Worthing, *A Review of Cost Estimation in New Technologies: Implications for Energy Process Plants*, Santa Monica, Calif.: RAND Corporation, R-2481-DOE, 1979. As of April 10, 2007:
http://www.rand.org/pubs/reports/R2481/

Merrow, Edward W., Kenneth E. Phillips, and Christopher W. Myers, *Understanding Cost Growth and Performance Shortfalls in Pioneer Process Plants*, Santa Monica, Calif.: RAND Corporation, R-2569-DOE, 1981. As of April 10, 2007:
http://www.rand.org/pubs/reports/R2569/

Office of the Secretary of Defense, *Selected Acquisition Report, Program: Javelin (AAWS-M)*, Washington, D.C.: Office of the Secretary of Defense, December 1989.

———, *Selected Acquisition Report, Program: AAWS-M*, Washington, D.C.: Office of the Secretary of Defense, December 31, 1990, p. 3.

———, *Selected Acquisition Report, Program: Javelin (AAWS-M)*, Washington, D.C.: Office of the Secretary of Defense, December 31, 1991, pp. 3–4.

———, *Selected Acquisition Report, Program: Javelin (AAWS-M)*, Washington, D.C.: Office of the Secretary of Defense, December 1993.

———, *Selected Acquisition Report, Program: CH-47F*, Washington, D.C.: Office of the Secretary of Defense, December 2001.

Office of the Under Secretary of Defense (Acquisition, Technology, and Logistics), *Acquisition Reporting Overview and Unit Cost Reporting/Nunn-McCurdy Breaches*, Briefing, May 2003.

"On Target With Cost," *Defense News*, September 13, 1999, p. 14.

Phillips, R., *An Analysis of Aircraft Weapon Systems Cost Growth and Implementation of Acquisition Reform Initiatives Using a Hybrid Adjusted Cost Growth Model*, Master's thesis, Wright-Patterson AFB, Ohio: Air Force Institute of Technology, A990324, March 2004.

Stark, F. T., "Why Military Airplanes Cost So Much and What Can Be Done About It," *Air University Review*, November–December 1973.

Tyson, Karen W., J. Richard Nelson, Neang I. Om, and Paul R. Palmer, *Acquiring Major Systems: Cost and Schedule Trends and Acquisition Initiative Effectiveness*, Alexandria, Va.: Institute for Defense Analysis, IDA Paper P-2201, March 1989.

Tyson, Karen W., Neang I. Om, D. Calvin Gogerty, J. Richard Nelson, and Daniel M. Utech, *The Effects of Management Initiatives on the Costs and Schedules of Defense Acquisition Programs, Vol. 1, Main Report*, Alexandria, Va.: Institute for Defense Analysis, IDA Paper P-2722, November 1992.

U.S. Air Force Systems Command, *The Affordable Acquisition Approach Study*, Andrews AFB, Md., February 1982.

U.S. General Accounting Office, *High-Risk Series. Defense Weapons System Acquisition Report*, Washington, D.C., GAO/HR-93-7, 1992.

———, *Future Years Defense Program: Optimistic Estimates Lead to Billions in Overprogramming*, Washington, D.C., GAO/NSIAD-94-210, July 29, 1994.

Wong, Carolyn, and Jeffrey Drezner, *Factors Affecting Cost Risk in Weapon System Acquisition*, Santa Monica, Calif.: RAND Corporation, WD-4561-AF, September 1989.

Wong, Carolyn, Jeff Drezner, and Ron Hess, "Findings from a Literature Survey on Cost Uncertainty," unpublished.